E. Rudolph   I.-O. Stamatescu (Eds.)

Philosophy, Mathematics and Modern Physics
A Dialogue

E. Rudolph   I.-O. Stamatescu (Eds.)

Philosophy, Mathematics and Modern Physics

A Dialogue

Springer-Verlag
Berlin Heidelberg New York
London Paris Tokyo
Hong Kong Barcelona
Budapest

Professor Dr. Enno Rudolph
Professor Dr. Ion-Olimpiu Stamatescu

Forschungsstätte der Evangelischen Studiengemeinschaft, Schmeilweg 5
D-69118 Heidelberg, Germany

With 10 Figures

ISBN-13: 978-3-642-78810-9   e-ISBN-13: 978-3-642-78808-6
DOI: 10.1007/978-3-642-78808-6

CIP data applied for

Cover Design: E. Kirchner, Heidelberg

SPIN: 10134403   55/3140 - 5 4 3 2 1 0 - Printed on acid-free paper

Preface

This volume represents an attempt to intensify the renewed dialogue between physics and philosophy brought forth by the development of quantum theory and to contribute in this way to its consolidation. The two introductory texts, conceived as dialogues themselves, open some of the questions and illustrate the discussion form which served as starting point for the approach followed in the enterprise documented in this book.

Many of the ideas presented here emerged from a series of interdisciplinary workshops devoted to philosophy and natural sciences and organized by the "Protestant Institute for Interdisciplinary Research" (FEST, Heidelberg). In the selection of contributions included here the authors are concerned with themes and questions arising from the basic problems of modern physics and its corresponding mathematical setting – problems which could not be treated without resorting to both the classical and the modern philosophy of science.

The framework for this project was developed following extensive discussions with Carl Friedrich von Weizsäcker. The project's organizers, who are also the editors of the present volume, drew essential inspiration and guidance from his pioneering work in this field and would like to take this opportunity to acknowledge his valuable influence.

The editors should like to express their appreciation to Springer Verlag for its open-minded interest in taking this volume into its publication program. Special thanks are due to Professor Wolf Beiglböck for his interest and expert councelling and to Dr. Angela Lahee and Mr. Andrew Ross for competent and invaluable active help in elaborating the manuscript and correcting the English. We should also like to thank Mrs. Franziska Strohmaier, Mrs. Ingrid Sproll and especially Mr. Tilman Bohn for their dedicated participation in the technical preparation of the manuscript.

Heidelberg   *E. Rudolph*
January 1994   *I.-O. Stamatescu*

Contents

Introduction

On the Dialogue Between Physics and Philosophy

Enno Rudolph and Ion-Olimpiu Stamatescu

FESt, Schmeilweg 5, D-69118 Heidelberg, Germany

1. What Have Physics and Philosophy to Tell Each Other?

"Die aristotelische 'Physik' ist das verborgene Grundbuch der abendländischen Philosophie" (Heidegger). This sentence refers to a work and a time, in which physics was a philosophical discipline and nature an object of philosophy. Accordingly, the philosophers of Greek Antiquity were just as much concerned with metaphysics as with the study of nature. This was also true in the time of the presocratian: Thales, Heraklit, Anaximander, Demokrit, Leukipp or Empedokles; later Platon, Aristoteles, Theophrast and finally Epikur. This is even more true of the enduring influence of this epoch extending until the early modern time. The research of Galilei and Newton, as well as the natural philosophy of Leibniz or Descartes, still represent the indivisible unity of natural science and philosophy, which was broken only when the practical success of empirical science led to a stronger emancipation from the philosophy, and the latter, as a reaction, tried desperately to react to this with highly systematized and organized systems of metaphysics (including metaphysics of nature).

Already the philosophy of Kant, which was at first seen as being of epochal importance for the development of the natural science, could soon no longer claim to be acknowledged as a foundation theory for the latter. The consistent development of the splitting between humanities and natural science, particularly between physics and philosophy, determines even today the research and discussion in these disciplines and between these disciplines. In the 19th century no philosophical theory which took nature as its primary object succeeded in influencing or making any impression on natural science. Such efforts by Schelling and by Hegel failed grandiosely. Meanwhile it appears that the complexity of the problems, both in physics and in philosophy, forbid even the smallest attempt to communicate between the two. Especially the dominant philosophy in Germany in the 1930s, the philosophy of Martin Heidegger, has established an understanding of science inside humanities which, under the guide line of renewed hermeneutic, is defined through a sharp differentiation from the natural sciences – in fact, by the claim of the impossibility of a dialogue between these two sciences. Natural science on the other side, whose language became just as ununderstandable for philosophy as the latter is for natural science, has itself retreated onto the terrain of mathematical logic, which it uses as a sufficient criterium for theory building on the basis of experimental results.

A new situation appears, however, with the beginning of quantum physics, which raises, inside the paradigms of physics, fundamental questions, such as those concerning the cognizability of nature, the objectivity of tangible objects, etc., questions which seek philosophical answers. Not only physicists but also biologists, neurophysiologists and biochemists turn more and more to philosophy to ask such questions, bringing a new spirit into the great tradition of the last century. From the question, are we at all able to establish objective laws of nature, to the questions, is everything we call "living nature" but the result of an increasingly complex selforganization process of the matter – there appears to be present in natural science a continuous and deep need for philosophical enquiry, which not only provokes philosophers to a dialogue, but also makes them curious and interested. The hermeneutic philosophy of the 20th century on the one hand, and the positivist school on the other – which is, however, more oriented towards the empirical science paradigms – have become weary of being concerned with themselves. But particularly those philosophers who are influenced by the traditional epistemology and traditional metaphysics, even more than the two latter schools, feel inclined to deal with the question coming from natural sciences in a constructive way. In addition to the awareness of a crisis in natural science, there came also the ecological crisis to make philosophy aware of the fact that the question of truth – which embodies its abstract interest – cannot be separated from that of the concept and the definition of nature and life. And there is a surprising agreement between natural science and philosophy on the necessity to clarify these questions before one can decide in what sense one can prescribe laws to nature (laws which it probably does not have "by itself"), and what kind of validity these laws may claim.

These observations can be interpreted as symptomatic of an increasing awareness of a kind of crisis in fundamental research in natural science – a need to restore and pursue a dialogue which became silent long ago. The question, if natural science and humanities have something to say to each other, especially physics and philosophy, is thereby already positively answered: at least this seems to be true inasmuch both sides have explicit expectations of each other. But at the same time there became problematically acute other questions: in which language should this dialogue take place, what common tasks have to be solved and, finally, what purpose should this dialogue necessarily and usefully serve?

The problem of finding a common language is the most difficult. Physics "speaks" mathematics (abstractly) or "every-day language" (concretely). Philosophy "speaks" conceptual analytically (concrete) or synthetic-speculatively (abstract). One cannot distribute exclusively the alternative concrete/abstract over the two disciplines. The two languages have become so autonomous, that a direct dialogue rightly seems impossible. To speak to each other one must learn a new language, with the special, additional difficulty, that this language does not yet exist (and perhaps cannot exist, after all, as a new metalanguage). But it is just this fact which forces us to take the risk of trying the experiment of learning to talk by talking – according to classical hermeneutic, the starting point of every dialogue. To try and to exercise such a dialogue is the first and most fundamental interest of the project "Philosophy and Physics" which runs since nearly two years at FESt and finds its expression in a series of very intensive colloquiums, the "Philosophy-and-Physics-Workshops".

By interesting both sides, the topics of discussion at these workshops could help to reveal the sought-for form of the dialogue. They represent fundamental questions: thus for instance, the question "what is nature?" – is this something given, mathematically describable and thereby evading more and more an empirical description and conceptual "visualization"? Does "nature" hold as a valid name for a world which can be objectively experienced by all humans? Do humans belong to it the same way as do animals, plants, stones or atoms? Does it have its own rules, or do we try to impose our control over it by prescribing what it has to accept as rules? To find this out there are not only the natural scientists who must force themselves to think in fundamental terms over definition and meaning of their object – but also the philosophers must let themselves be told by the former what these actually accomplish, when they claim they contribute to the enlargement and progress of the human knowledge. And to do that, the philosophers must themselves learn to put questions again.

Crises can lead to productive revolutions – at least in sciences. Thomas Kuhn has made here important and well defined proposals in his fundamental book on the history of the scientific revolutions. But crises can also lead to destruction. In natural sciences it is possible that theoretical crises acquire practical significance when the object which has been lost from view loses its defense, gets out of control and is destroyed. They can also prevent the earlier paradigms, victims of the revolution, to act as a corrective. To diagnose the possibility of such developments one needs a reflective act which forces the natural science to philosophy and which forces the philosophy to adopt again its old and genuine aristotelian object: the unity of nature.

Of course, the situation has become much more difficult in the meantime, not only because of the reasons described above and pertaining to the increasing autonomy of the sciences, but especially because of the distribution of competence, and the high complexity, both in physics and in philosophy. Also, one lacks a place and framework in which this dialogue can be exercised, without disrupting or endangering concentration on the disciplinary work.

The FESt has developed such a framework by bringing together twice a year philosophers, natural scientists (mostly physicists) as well as theologians to talk and to conduct discussion on the questions exposed above.

2. On the Character of the Questions in the Philosophy-Physics Dialogue

On the philosophy side, the traditional fields – natural philosophy, epistemology or even antique and early-modern metaphysics – appear to be more adequate partners in the discussion with modern physics, then e.g. the theory of science or the language-analytical philosophy. The latter have distanced themselves and became autonomous to such a degree, that in their field of competence neither the (originally) ontological question on the "Being of nature" nor the (in the narrower sense) hermeneutic one on the "Meaning of the Being of nature" are accepted any longer as being qualified for discussion.

However, these are precisely the questions which characterize the beginning of the dialogue between philosophy and physics and they soon lead to the constructive question, whether an "interdisciplinary hermeneutic" may be achieved which should be able to conduct the hermeneutic debate of the last decades out of its self imposed limitation to human sciences. The concepts of "Sense" and "Meaning" – very subtle philosophical constructs indeed – will thereby be subject to test on their productiveness, in particular for the domain of natural science. But what do we call "Sense" or "Meaning" in a natural science 'sense'? Do we give "Sense" to a state of affairs or to a theory describing it by – freely following Hume – just getting used to the committing character of some interpretation of this state of affairs? Is "Sense" hence only a convention? Or does it emerge out of a series of connections, made by us in a more or less conscious way on the basis of an undefined multitude of observations, and in accordance with pre-given concepts? The historian of philosophy will notice that this alternative can be modeled on the dispute between David Hume and Immanuel Kant. This is therefore an epistemological question so far. But it can also be enlarged and acquire an ontological dimension, and then it turns into the question, whether "Sense" should refer to a (to us) external order, called nature, or it is only the trace of structure imprinted by us on an amorphous external world. Finally, there would also be the question, whether "Sense" is the result of a reflection on the logic of our thinking, or it appears as an even condition of our thinking and hence of our concept building itself. Similar questions concern the concept of "Meaning", namely both "Meaning" of the concepts we use and "Meaning" of the already in terpreted connections.

The principal difficulty of these questions consists primarily in the fact, that they unavoidably move in a circle between Concept and Being, a circle which e.g. led E. Cassirer to define Meaning as "die Einheit der Beziehung, kraft deren ein Mannigfaltiges sich als innerlich zusammengehörig bestimmt";

since " 'Begreifen' und 'Beziehen' erweisen sich der schärferen logischen und erkenntniskritischen Analyse überall als Korrelata, als echte Wechselbegriffe." One should not forget thereby, that words like "Meaning" and "Sense" often play the role of world picture categories with unseparable psychological components. In an interdisciplinary dialogue between physics and philosophy one must first try to envisage the frame in which the question over "Sense" or "Meaning" will be put, and then build up a connection between the various problem formulations, as they are found for instance between the previously mentioned questions on the Being of the nature and on the Sense of its Being. In the context of the self-image of physics probably everybody will agree, and in a rather wide sense, with Heisenberg's sentence: "Natural science does not simply describe and explain nature; it is part of the interplay between nature and ourselves. It describes nature as exposed to our method of questioning." Hence, in connection with knowledge (of nature), Meaning and Sense are generated in a conversation which has to bring forth certain conceptual connections and interpretations. It is remarkable, that Cassirer's "Wechselbegriffe" (exchange concepts) and Heisenberg's "interplay" describe a coercion to dialogue between nature and subject, which suggests a simultaneous discussion of the corresponding approaches of these two thinkers. This might be a model for the sought for dialogue between physics and philosophy.

Let us consider for illustration a question which one loves to put nowadays, namely: do physicists really believe that the universe has a finite age and extension? Now this question accommodates in fact more points of view. It is first a physical question, that is, a question concerning our experimental (and observational) data and the theories which deal with them. Here one can answer, yes, up to a number of subtleties, on one hand, and of uncertainties, on the other hand, we recognize a finite, unlimited universe (at least in our most accepted – dominant – theory). But this is also an epistemological question, a question of reference and interpretation, which is in fact indisolubly related with the physical one; and here the above mentioned subtleties and uncertainties become very important. The subtleties tell us, first that physical space and time *are defined* by the universe, since the concept itself of the physical universe implies that it contains *all* physical things – this is sometimes called the "content principle" in cosmology; the universe is the primordial concept, while space and time are subordinate concepts. One should not think of an absolute space and time in which the universe is embedded (and takes, possibly, a finite part) – this would reverse the above hierarchy. And secondly, that the finiteness shows up in the frame of an extrapolation of our usual concepts of space and time beyond the point where their present understanding loses any meaning, any possibility of reference inside the conceptual net in which they were originally born and in which they are active. And the uncertainties have to remind us of the hypothetical character inherent in our theories, the more so that the present stage is surely not satisfactory (we lack a consistent quantum theory of space-time). Finally, it is here also a question of interpreting knowledge into a scheme of self-understanding, of intentions and of believe. Here the augustinian understanding that time itself is created together with the universe and that it is therefore meaningless to ask about "before" and to wonder about the finite age of the latterrepresents a spiritual attitude which appears consistent with the physical picture we started with, but which is in fact a statement about the historical time and the timelessness of God (concerning the question "did time exist before creation - of time, of the world -?" see Augustinus, Confessiones, 11th Book). These various "answers" do not seem to differ in their statements themselves, but the more they appear similar the more they differ in their roles in the meaning-giving activity. They need therefore to retain a certain autonomy, such that, for instance, one may find an augustinian point of view compatible both with a Big Bang and with a steady state model, while leaving experience to chose between the latter (what in fact has been the case). But they are also not completely independent of each other, as was clear above, e.g., in stating the "physical" and the "epistemological" points of view. It is the task of the dialogue between physics, philosophy and theology to clarify these different validation schemes in comparison to each other and to promote a real understanding and flow of ideas, avoiding ambiguities.

Under the official physical disciplines it is theoretical physics which leads the conversation. Theoretical physics is not an autonomous science: it gains its objects, concepts and methods in interaction with other disciplines, whose research results are structured by it, ordered in a relatively independent scientific connections frame and thereby also transformed. A special interdependence exists between theoretical physics on the one side and experimental physics, phenomenology and mathematics on the other. The experiment must produce classifiable and reproducible "experiences". These two requirements – classifiability and reproducibility – suggest the possibility of a feedback, which Einstein pointedly characterized by the sentence "die Theorie sagt erst, was man messen kann". The theory seems thus to be a corset, a frame which can never be shattered. There may well be, however, explicit inner inconsistencies in some physical theories – such as the so-called ultraviolet catastrophe in the non-quantum physics, which shows up in the attempt to put forth a classical theory of radiation for continuous media. Or there may be contradictions between observations and the given theory, such as the stability of atoms – inexplicable in classical electrodynamics. Such contradictions seem to point to a "reality" which "resists" both the experimenting and the theorizing subject. Indeed, it is not clear why the observed stability of matter, say, should imply an inner inconsistency of the classical theory – at least as long as one does not resort to a principle which has been commonly termed "anthropic" nowadays. But one also cannot reject the argument that these contradictions or resistances would have turned up as inner inconsistencies in the (vain) attempt to obtain an overriding and more general theory on the basis of classical physics. Even then, however, experiment should retain its importance as troublemaker in exhibiting these inconsistencies and hence motivating us to search for the correct general theory explaining those experimental results (that is, predicting them correctly as a question of inner consistency!).

Mathematics allows physics to build up a consistent conceptual scheme, in which concepts define themselves together with their relations – Heisenberg speaks of physical theories as "closed systems of concepts". The development of physics is essentially influenced by that of mathematics, a fact which, however, cannot be correctly described without considering also the reverse influence, from physics to mathematics. If we disregard indeterminacies in mathematics itself and also do not take into account that the development of the mathematical scheme is itself a highly nontrivial, and often very difficult process, then we can consider all physical theories as (at least potentially) closed and secure as concerns their mathematical scheme. It is therefore unclear where might be the origin of the above mentioned inner inconsistencies. In fact, these inconsistencies do not show up primarily in the mathematical scheme of the theory, but in the relation between this scheme on the one hand and its reference to certain classes of phenomena on the other. Such is the case in the application of classical radiation theory to systems with infinitely many degrees of freedom (continuum). They are, however, still to be reckoned as inner inconsistencies of the theory, since the physical theory is not characterized only by its mathematical scheme but also by its claim to be consistently applicable to certain well defined classes of phenomena – namely, defined in the frame of the theory itself and not by extratheoretical ad hoc criteria.

The arena of the phenomenal reference of theories is given by phenomenology – which, as physical discipline, is concerned with the structuring of experience. Rules will be established here, which will serve as a starting point for the mathematical abstraction and construction processes, and predictions from theories will be interpreted here, for the purpose of experimental tests. Phenomenology influences, directly, through its tendency to model and to provide pictures, the concept building of theoretical physics. One can ask why we make a distinction at all here, since officially phenomenology is just part of theoretical physics. The reason for this differentiation is that phenomenology does produce laws but does not build up theories. A physical theory is never a simple structuring of experience. A theory proceeds from a set of relevant experiments or observations, develops in a field of already existing knowledge and uses a mathematical scheme which it finds or adapts, in order to build up an organic whole – "a closed system of concepts". In this way not only will predictions be obtained for new experiences, but also the general theoretical structure will be enriched by new relations, refinements and even new concepts. One of the most fundamental epistemological questions which characterizes the physics-philosophy dialogue is then: according to what criterion will new concepts be generated and accepted and how is this process connected to the question raised above on the "Sense" and "Meaning" of knowledge? These and other related questions allow us to introduce a structure in this dialogue and to provide it with a frame complementary to that of the theory of science, in that its primary flow is not from philosophy to science but from science to philosophy.

3. An Experiment in the Dialogue Between Physics, Mathematics and Philosophy: the "Philosophy and Physics Workshops" of FESt

The "Philosophy and Physics Workshops" of FESt represent an experiment in the dialogue philosophy – natural sciences. The setting of this experiment is determined by the readiness of the physicists and of the philosophers to enter into a discussion, in which both the purely physical and the purely philosophical arguments can be much more detailed and specific than is usually considered acceptable for an interdisciplinary discussion. Although dangerous, this approach offers the chance of discussing controversial questions at the level of specificity required by the argument. The profit from this approach lies in the fact that the contradictions existing inside each field are uncovered, requiring simultaneously to be explained "to the outside". This offers the possibility of witnessing the research situation in the "other" field, and also to becoming aware of the reach of the arguments of one's own field from the point of view of the other. It is in the interest of the common discussion to prevent this from becoming too technical and decaying into a multitude of partial aspects. Our questions themselves do not proceed from a pre-given, clear concept of the "unity of physics" (C.F. von Weizsäcker) – but, of course, they do assume the possibility of such a "unity" as a "regulative idea". The workshops represent also in this respect an "experiment" in the physical-philosophical dialogue, since the common interest guiding the accumulation of knowledge is to find a frame for a theory of nature, which can only be co-operatively formulated in a physical and philosophical collaboration.

The themes treated up to now have been chosen according to this point of view. The series of Philosophy and Physics Workshops opened in spring 1989 with a general discussion, which treated aspects of what may be called the "fight over the universals" among physicists. This "fight" is triggered – to put it tersely – by the question: is there (and are we concerned with) the nature "of things", or only (with) our knowledge of it? The contributions to this workshops have been published as number 1, volume 27 (1990) of the journal *Philosophia Naturalis*. A second workshop followed in fall 1989, which was more details-oriented and had as working title "The concept of nature and the meaning of reality in modern physics". Its contributions appeared as number 1, volume 28 (1991) of the same journal. For the realization of our program it was important to document the venture of this interdisciplinary experiment in a journal with high reputation in both disciplines and it is the first time that *Philosophia Naturalis* has devoted a whole number to the proceedings of a conference. The next workshop, June 1990, was again concerned

with a more general theme: "Cosmology". The the fourth workshop followed
in February 1991, under the working title "Physical and philosophical aspects
of our understanding of space and time". The next workshops (February 1992
and February 1993) had the following themes: "Theological and physical cosmology" (5th) and "The mathematics of physics" (6th).

Before proceeding to the presentation of the contributions collected in the
present volume we shall briefly recall the discussion pursued at the first two
workshops, to establish the connection. In fact, these discussions are substan-
tially more interrelated than may be apparent from the above of succession
themes, and they also do not, as a rule, proceed in closed steps.

The themes discussed at the first two workshops can be divided into three
fields:

1. Foundation problems in modern physics
2. The question of reality in quantum mechanics
3. Objectivity and natural law.

We shall consider two special aspects of this discussion.

a) Quantum Mechanics and the Classical World

Quantum mechanics describes phenomena for which no classical pictures exist:
such a phenomenon is for instance a microscopic particle, for which position
and speed can never be simultaneously sharply defined. On the other hand,
the standard quantum mechanical description of these phenomena implies a
"classical apparatus", a device like a photoplate or a particle counter, which
of course will always produce sharp results (inside classically understandable
experimental errors): we do not even know how an "unsharp event" may look
(what is the unique event "here *or* there" if we speak of spots on a plate?).
The fundamental "coherent ambiguity" of the quantum mechanical state – as
implied by the superposition principle and as shown, e.g., by the interference
pattern of the well known double slit experiment – is resolved by the classical
apparatus into sharp results and classical probabilities: after the passage of a
particle through, say, the experimental arrangement of the double slit type,
only one (sharp) point will appear on the photoplate, just as a classical object
would produce. However, at each passage another point on the plate will be
"hit" and the accumulation of the points on the plate will be distributed
according to the probability density calculated by quantum mechanics. The
"measurement", whether it only means a change in our information structure
or it implies a real event ("collapse" of the wave function), is a nontrivial
process which is essential for the theory.

In this connection the question then arises as to the extent of this "duality". Is this restricted to the axiomatic framework of the theory, or do we
have to postulate an irreducible classical world next to the quantum one? Can
one explain quantum mechanically that, say, a tennis-ball always appears well
localized, that for it no interference effects can be proven? (if you find the

example not well chosen you may think of the classical piece of chalk instead - or of Schrödinger's cat). In fact, one has always assumed this to be the case – the ambition of physics has always been, however, to show *precisely* how an effect obtains in the framework of a theory, because this is what brings secure understanding of its emergence. The contributions of E. Joos, H.D. Zeh and C. Kiefer have shown how, for bodies with very many degrees of freedom (many variables – many parts, e.g. atoms), classical behavior can emerge as a result of countless and uncontrollable interactions with the environment. Although the state of the total system (body and environment) remains coherent, in accordance with the quantum mechanical description, the "practical" difficulty of controlling the environment leads to an effective loss of coherence as referred to the macroscopic body alone – no interference effects can ever be put into evidence for it and "for all practical purposes" it will always appear classically (the correlations are "delocalized", they are no longer accessible to the observer – assumed "local"). While this result by itself does not explain the measurement process, it is however very important for the consistency of the theory, since quantum mechanics as the "correct theory" should also explain the classical world, the phenomena which appear classically – otherwise one is forced to limit its domain of validity in an ad hoc way. Beyond its theoretical impact, this analysis is also epistemologically relevant, since it shows that the "quantum world" and the "classical world" are not disjunct and alien to each other but are appearances of one and the same world and can be related by a continuous transition.

b) Reality and Objectivity

To search for a reality "behind" the quantum mechanically best describable "object", the quantum state (the wave function), is very problematic. Due to the "coherent ambiguity" mentioned above, which arises because we can build a valid state by superposing states specified by different values of the same measurable property (observable), we do not even know how to characterize what we are looking for. But to endow the wave function itself with reality is also precarious, since we then have to live with its fundamental non-separability and hence reckon with non-local concepts: either we must envisage a non-local event, like the collapse or reduction of the wave function, which may run into conceptual trouble with special relativity, or we must imagine a restlessly splitting observer, or consciousness, a not necessarily convincing picture.

A similar situation holds for the concept of objectivity, since we cannot easily speak of objective properties if these cannot always be uniquely attributed. In a series of contributions (d'Espagnat, Neumann, Rudolph, Scheibe, Selleri, Stamatescu) this problem has been discussed in connection with the question on the definability of the concept of object, whereby various approaches, coming both from traditional philosophy (Leibniz, Hume, Kant) and from more recent philosophical research (Cassirer), have been analyzed from the point of view of their relevance for the concept of object in quantum mechanics. In particular the solution proposed by Cassirer concerning the epistemological foundation of quantum theory proved to be especially fruitful and of astonishing timeliness. The disappearance of the "substantial" object concept of classical physics was shown by Cassirer to have its philosophical roots in the philosophy of Leibniz. The mathematician, logician and philosopher Leibniz enters then in direct dialog with modern physics, even more than Kant and his epistemological a priori theory.

This short presentation can only serve recollection or illustrative purposes and to indicate the continuity of the discussion.

4. The Problems and Background of the Contributions to this Volume

The 3rd and 4th Philosophy and Physics Workshops ("Cosmology" and "Physical and Philosophical Aspects of our Understanding of Space and Time"), from which some of the articles in the present volume have been selected took place at FESt in June 1990 and February 1991, respectively. They not only build up a conceptual unit, but they also represent a new step in the dialogue, characterized by a deepening and sharpening of both the epistemological and the physical questions. The 5th workshop "Theological and physical cosmology", February 1992, introduces new questions concerning *Weltanschauung* and the interaction between science and theology. The 6th workshop, "The mathematics of physics", February 1993, opens the discussion about the fundamental epistemological question of the role of mathematics in physical science. Since this volume contains a selection of contributions to these workshops we shall proceed with some comments on their background. In connection with all these problems the physical discussion must observe the fact that here the frontiers of research are tackled, and the epistemological discussion must consider the problems of present day physics, account for their incentives and use their hints.

Natural philosophy in this century is defined among other things by two aspects: the "problem of reality" – such as it is raised especially in the context of the realism versus positivism argument in modern physics – and the "problem of objectivity", whose different treatments characterize the various epistemological *Ansätze* (transcendental method, analytical philosophy, evolutionary theory of knowledge, etc.). These two aspects are of course correlated, namely in different ways in the different philosophical systems. Also it becomes explicit here that philosophy of nature as philosophical discourse about nature cannot abstract from the epistemological questions. This therefore makes the bridge to the theory of science, to "protophysics", to "foundation research", etc. – even if these fields correspond to a further specification of the questions and therefore will remain outside our considerations here.

The theoretical and experimental results of recent years concerning basic questions from quantum theory (Bell's inequalities, decoherence, classical properties) have contributed to making precise the questions of relevance for epistemology and natural philosophy. But also beyond this, present day physics shows an exciting fullness of new aspects, which lead to both sharpening and widening of the discussion, since it seems that some of the basic questions of modern physics only now show their true depth; such are, for instance, questions about the space-time continuum, about cosmology, etc.

But what is "present day physics"? The picture of contemporary physics research seems characterized by an interplay between efficiency and fundamentality. Namely, there exist a number of quantum field theories which describe separately and with a certain success the various classes of elementary phenomena, that is, they achieve a certain level of "efficiency" in accommodating empirical data with a restricted number of free parameters. In the frame of these theories one arrives at a certain level of unification between these phenomena, hence at a certain "fundamentality level" in the reduction. There are many attempts to reach through further unifications to a more fundamental level. These unifications are, however, mostly too imprecise (either in their structure or in their predictions) and therefore they are not successful enough to reach high efficiency: they cannot really be tested yet. (This should not be interpreted as a criticism of the search for a fundamental description. But it means that apparently at present essential information – theoretical and experimental – is missing, which would have helped to turn speculations into a compulsory scheme. This is, for instance, the reason, why Heisenberg's program of searching for a unified field theory failed.) There are also many "effective" theories which describe subclasses of a class of phenomena by replacing the influence from the rest of the phenomena in that class with "effective" parameters, which are kept free at the theoretical level and must be fixed empirically. Although these models may be very successful, they again have a low efficiency, since they must use ad hoc assumptions to allow experience to fix quantities which should in fact be determined theoretically.

At each given moment, therefore, a maximum of the efficiency appears, represented through a more or less coherent set of models and theories and defining a certain depth of the "reduction process", of fundamentality. Although one usually speaks here of "standard" or "dominant" theories, these considerations overlap only partially with Kuhn's dominant paradigms. Physics proceeds by building hypotheses on the background of a certain theoretical understanding and empirical information and at the highest level of construction these hypotheses concern full theories. There is therefore a natural inertia appearent in this process, which is only increased by the fact that falsiability is not easily achievable, both theoretically and empirically: evaluation of the internal and factual contradictions and the competition between hypotheses are themselves part of the conceptual development. The notions of "success" and "efficiency" introduced above are therefore rather complex and difficult to define explicitly, they have however not an essentially arbitrary character but are – even if imperfect – tools of the argumentation.

The "dominant set" of theories of contemporary physics is defined through the classical general relativity theory (GRT) and the quantum field theories (QFT) of the standard model of elementary particles (SMEP: quantum chromodynamics, QCD, and the Glashow-Weinberg-Salam model of electroweak interactions, GWSM). It includes further a number of theories which are understood either as approximations and limiting cases or as generally valid theoretical schemes (mechanics, electrodynamics, statistical mechanics, quantum mechanics, thermodynamics). To describe the position of the dominant theories on our "efficiency diagram" we should notice first that the developments of the last decades hint at the search for a general, unified, parameter-free theory. Such a theory would contain (beyond the four "constants of nature" c, h, k_B and probably l_P) no parameter to be fixed empirically – in fact it should even predict the dimension of space-time. The concept of parameter-free theories arose in the context of the so-called renormalization program for asymptotic free theories (like QCD). The four constants of nature represent the freedom we have in the introduction of various quantities of empirical relevance. The first three can all be set to one, by which we just fix a system of units in which all physical quantities can be measured in units of (powers of) l_P. Thus a velocity of 0.1 means 1/10 of the velocity of light (i.e., 30 000 km/sec) and a temperature of $10^{-29} l_P^{-1}$ just this fraction of the inverse Planck length (amounting to around 1 K in usual units). This "commensurability" is based on the existence of physical phenomena, the theoretical descriptions of which involve transformations between different quantities. Thus for example the symmetry transformations of relativity theory imply a commensurability (not identification!) of space and time; statistical mechanics of ideal gases relates temperature and energy; the temperature of a black body is related by a universal law to the frequency spectrum of its radiation; both the Schwarzschild radius and the Compton wavelength – fundamental concepts in GRT and QFT – are defined in the frame of relations connecting energies and lengths and some kind of relation between them, as expected to be established in quantum gravity; all of which would introduce l_P as the fundamental dimensional unit. (We stress again that the quantities related in this way are not identified. We can meaningfully speak, for instance, of frequency on one hand and of temperature on the other hand in a lot more cases than those where we can speak of both together. The commensurability is based on the existence on a nonvoid set of cases where both concepts apply and can be related by a universal law: in this instance the black body radiation law.) The role of the four constants of nature is to allow a translation from the scale of fundamental phenomena to our scale by providing a "second", a "gram", a "kelvin" or a "meter". Their existence is relevant, while their values are only of interest for setting the human scale, but not for the structure of the theory.

A very different situation holds for "free parameters" (dimensionless numbers in the $c = h = k_B = 1$ system) which must be fixed from outside of the given theory. As long as such parameters must be fixed empirically one can ask whether there should not be a more powerful, fundamental theory which would fix them free of empirical information. Such is the case, for instance, with QCD, the theory of strong interactions, which fixes the masses, charges and further properties of all "hadrons", the particles entering all the nuclear phenomena (proton, neutron, mesons, etc. – previously thought as elementary, later found to show a structure at small distances, i.e., compositeness). These masses and further properties are no longer free parameters in QCD, in contradistinction with the masses of quarks, the components of hadrons and fundamental particles in QCD: presently empirically determined, the properties of the quarks should themselves result as predictions in the frame of a "Grand Unified Theory", as also, e.g., the electric charge and the lepton masses (electron, etc.). The picture one has developed in the last two decades is that the fundamental theory should belong to the class of "asymptotically free" theories, like QCD itself, and therefore should contain no coupling constant as free parameter. It would be therefore the parameter-free theory we mentioned above, the "theory of everything" (ToE). Such a theory is, however, not in sight, not even concerning the true unification of the electroweak and the strong interactions (the SMEP is put together out of QCD and GWSM as separate parts). The principal difficulty seems, however, to reside in the quantization of gravity. While one has some ideas about quantum gravitational effects, no consistent theory has been constructed yet. This difficulty influences also the related problem of quantum cosmology. Therefore also the problem of understanding space and time is held open, and this the more so while QFT hints at an intimate connection between space-time and interaction.

A further very important aspect comes into the game here, namely, that the abstraction level necessary for the establishing modern theories leads to a kind of phenomenological opacity. Even when we succeed in writing down such theories in a few axioms and formulae, starting from a number of general principles and empirical observations and using well defined symbolic structures (in itself a highly nontrivial problem, in a strong sense rarely achieved), it is very difficult to derive (calculate) empirically testable predictions from them. Such predictions are, however, absolutely necessary in order both to test and to understand (interpret) the theory. Therefore many of these theories are still sealed books. It is a very important part of the research effort in modern physics just to open and understand these books.

Finally, the deeper we go into the direction of fundamental processes, the farther we depart from the human scale – both toward small scales and toward large scales – and the more difficult is to obtain precise observational and experimental data. But there is no indication of some kind of convergence in the fullness of models and speculations in present research toward some unique scheme (a pious hope regularly deceived), although some general patterns may be observed. Hence there is no way to tell how decissive is the empirical information we miss – not even whether this game will ever end, whether we shall ever have more than an "effective ToE", reaching up to just before the observational horizon and liable even to strong changes in its basic structure each time the horizon reaches a new scale.

The present volume includes a selection of contributions to the 3rd, 4th and 6th workshop, which by their argumentative or even provocative character seemed adequate to sustain a vivid dialogue. Here both the more or less secured knowledge and the discrepancies and uncertainties mentioned above will become apparent. The more descriptive contributions from the 3rd, 4th, 5th and 6th workshops have been collected in a separate volume appearing in the series "Texte und Materialien" of FESt. We strongly recommend this accompanying volume for background discussions and information (see announcement at the end of this volume).

We have grouped the articles of the present volume in two parts, with only limited consideration for the original ordering. The first part, *Space and time, cosmology and quantum theory*, is opened by a very sharp question about the very possibility of founding our notions of space and time on conscious events (R. Penrose: *Is conscious awareness consistent with space-time descriptions?*). There follows a discussion of the problems encountered in relating the "intuitive" and the abstract, physical concepts of space and time (F. Lurçat: *Space and time: a privileged ground for misunderstandings between physics and philosophy*). The third article (I.O. Stamatescu: *On renormalization in quantum field theory and the structure of space-time*) discusses the change in the physical concept of the space-time continuum, from a description basis toward a dynamical element of the theory. There follows an argument extrapolating from the well known discussion about the "reality problem" in quantum theory toward an absolute consciousness (E. Squires: *Quantum theory – a window to the world beyond physics*). The conceptual necessity of putting the problem of quantum gravity in connection both with quantum theory and with the theory of space and time is discussed in the next article (C. Kiefer: *Quantum cosmology and the emergence of a classical world*). An illustration of the application of leibnizian optimization arguments, here understood as "maximalization of complexity", is described in the following article (J.B. Barbour: *On the origin of structure in the universe*). The standard model of cosmology is subjected to a critical analysis in the next contribution (H. Arp: *Galaxy creation in a non-big bang universe*) and the possibility for alternative pictures is presented. Finally, the last article of this part discusses the general problems of an "extrapolation science" such as cosmology and the basis of the standard cosmological model (H.F. Goenner: *What kind of natural science is cosmology?*).

The second part "Philosophical Concepts and the Mathematics of Physics" opens with an article dealing directly with the problem of reference, in the frame of an argument about concept formation and the role of a priorieity (F. Mühlhölzer: *On the assumption that our concepts structure the material of our experience*). The next article (E. Scheibe: *The mathematical overdetermination of physics*) introduces a fundamental question concerning the epistemological role of mathematics in physical science: the problem of grasping the knowledge building function of the mathematical structures in theoretical physics. The third article (K. Fredenhagen, *The mathematical frame of quantum field theory*) contains a presentation of the mathematical setting of our fundamental theories and its role in providing new connections and points of view. The next article (G. Münster: *The role of mathematics in contemporary theoretical physics*) introduces and discusses the mathematical aspects of modern physical theories, their epistemological implications and their significance in the process of frontier research. One of these aspects is taken over for a general consideration in the following article (E. Scheibe: *A most general principle of invariance*). The next contribution (G. Prauss: *Kant and the straight biangle*) is an analysis in the frame of the kantian conception of geometry. The volume closes with a discussion about the possibility of founding the concept of function (as it has been developed in modern philosophy especially by Ernst Cassirer) in the leibnizian thinking (E. Rudolph: *Substance as function*).

Acknowledgement

The authors are very indebted to Al Actor for improving the English of this article.

Questions Concerning Theory and Experience and the Role of Mathematics in Physical Science

Ion-Olimpiu Stamatescu and Heinz Wismann

FESt, Schmeilweg 5, D-69118 Heidelberg, Germany

The following considerations do not represent a well defined argument but rather a number of questions and remarks. They owe much to comments from Enno Rudolph.

1. One feels that a discussion about the role of mathematics in physical science involves the problem of referring theoretical concepts to physical experience, and that it does so in various ways. On the one hand there is the question of freedom or necessity and of redundancies in introducing quantities or theoretical "objects" in the frame of some mathematical scheme pertaining to a given physical problem. On the other hand there is the question of the role of consistency argumentation versus empirical determinacy in developing theoretical models. There is also the question of invariants in the theoretical schemes and their possible understanding as a priori. In connection with present day physics research, which is dominated by the regulative idea of a "theory of everything" (ToE) and which has a high level of abstraction characterized by a wide use of higher geometry and algebra to construct speculations toward the ToE, the reference question becomes primary.

2. For an adequate description and analysis of the interplay and mutual reference of theoretical prerequisites and empirical facts [empirisch erfahrbare Gegebenheiten] it may be interesting to look for a generalization of the model of the relations between categories and empirical representations [Vorstellungen] which has been developed by Kant in his transcendental deduction [transzendentale Deduktion] and try to see to what extent such an approach may provide a useful scheme for the epistemology of modern physics. We shall not attempt to do this job here, but we suggest using a discussion about this possibility in connection with some questions about theory and experiment and the role of mathematics in physics.

3. Staying first with this model we are urged to differentiate carefully between transcendental elements – preconditions for the possibility of experiences [Bedingungen der Möglichkeit der Erfahrungen] – and empirical elements. We shall ask later on whether and how this is possible; for the time being we only notice that in this differentiation we must avoid dogmatically hypostatizing either of them. Reality [Wirklichkeit] is neither that which can be empirically experienced, nor something residing fully in the consistency of the theory. Reality is that, what can be experienced in the frame of the theory (Einstein would say, the theory tells what can be measured).

4. We may notice that the primary motivation for the transcendental method simply relies on the observation that all reference to an experience implies theory (concepts, schemes, etc.) hence all direct experience depends on a transcendental process of meaning giving. The "critical approach" is the reflection about this process. Hume anchors this process of meaning giving in a type of certainty that no longer can be described in philosophical terms but only in psychological terms (custom and belief). Kant considers that the (observed) stability of this process must imply a more fundamental scheme. However, he cannot find this scheme in experience. It may be interesting to quote Kant's justification for the a prioricity of space as an example for his reasoning: "Der Raum ist kein empirischer Begriff ... Denn damit gewisse Empfindungen auf etwas außer mir bezogen werden ... imgleichen damit ich sie als außer und neben einander, mithin nicht bloß verschieden, sondern als in verschiedenen Orten vorstellen könne, dazu muß die Vorstellung des Raumes schon zum Grunde liegen. ... Der Raum ist eine notwendige Vorstellung, a priori, die allen äußeren Anschauungen zum Grunde liegt. ... Auf diese Notwendigkeit a priori grundet die apodiktische Gewißheit aller geometrischen Grundsätze, und die Möglichkeit ihrer Konstruktion a priori. Wäre nämlich diese Vorstellung des Raumes ein a posteriori erworbener Begriff, der aus der allgemeinen äußeren Erfahrung geschöpft wäre, so würden die ersten Grundsätze der mathematischen Bestimmung nichts als Wahrnemungen sein. Sie hätten also alle Zufälligkeit der Wahrnehmung, und es wäre eben nicht notwendig, daß zwischen zween Punkten nur eine gerade Linie sei, sondern die Erfahrung würde es so jederzeit lehren. Was von Erfahrung entlehnt ist, hat auch nur komparative Allgemeinheit, nämlich durch Induktion. Man würde also nur sagen können, so viel zur Zeit noch bemerkt worden, ist kein Raum gefunden worden, der mehr als drei Abmessungen hätte." [1] This small text can be used

---

[1] In our ad hoc translation: "Space is not an empirical concept ... since for some sensations to refer to something outside myself ... such that I can represent them as external and beside each other, thereby not as just different, but at different places, the representation of space must already stay at the basis of this referring. ... Space is a necessary representation a priori which stays at the basis of all external intuition. On this necessity a priori is based the apodictic certainty of all geometric principles, as well as the possibility of their construction a priori.

in fact to show both the strength of Kant's argument and its difficulty – to which the critical discussion will connect later on (see point 6 below).

5. Kant puts at the basis of his reflection the synthetic unity of apperception, the transcendental unity of self-consciousness [Selbstbewußtsein], as a faculty of producing the representation "I think" which must be able to accompany any other representation and hence realize the synthesis of the representations obtained from the diversity confronting our intuition [Anschauung]. From this he then deduces his scheme of "concepts of the pure reason" [reine Vernunftbegriffe], that is to say, of the categories [Kategorien], among others, the causality principle. Together with the two pure forms of intuition [Anschauung] – space and time – they build his scheme of the synthetic a priori. The objectivity of a priori knowledge comes from it becoming part of the objects of experience since it is the precondition for experience.

6. If we now look for the possibility of generalizations of the Kantian model we first note two of the points of criticism (see, e.g., Jammer 1988). The first is the rigidity of the a priori scheme, the second is the apparently subordinate or at least unclear role of the empirical element (his metaphor of the reason as "mounted judge" is illustrative).

Already in classical physics the first point brought forth contradictions, such that, e.g., Helmholtz would say: "Kants Lehre von den a priori gegebenen Formen der Anschauung ist ein sehr glücklicher und klarer Ausdruck des Sachverhältnisses; aber diese Formen müssen inhaltsleer und frei genug sein, um jeden Inhalt, der überhaupt in die betreffende Form der Wahrnehmung eintreten kann, aufzunehmen."[2].

Now, in classical physics, it was possible to introduce instead the "reality hypothesis", again according to Helmholtz: "Die realistische Hypothese ... sieht als unabhängig von unserem Vorstellen bestehend an, was sich in täglicher Wahrnehmung so zu bewähren scheint: die materielle Welt außer uns. Unzweifelhaft ist die realistische Hypothese die einfachste, die wir bilden

---

Were the representation of space an a posteriori obtained concept, extracted from general external experiences, the first principles of mathematical determination [Bestimmung] would be nothing else but perceptions. They would hence have all the randomness of perception, and it would not be necessary that between two points there be only one straight line, but the experience would tell us this all the time. What is taken over from experience has only comparative generality, namely by induction. One could hence only say, as far as we have observed yet, no space has been found having more than three dimensions." Kant (1781), B39.

[2] In our ad hoc translation: "Kant's teaching of the a priori given forms of intuition is a very lucky and clear expression of the matters; but these forms must be empty of content and free enough to be able to take over every content which may be encountered in the corresponding form of intuition." Helmholtz (1878), 299

können, geprüft und bestätigt in außerordentlich weiten Kreisen der Anwendung, scharf definiert in allen Einzelbestimmungen und deshalb außerordentlich brauchbar und fruchtbar für das Handeln. Das Gesetzliche in unseren Empfindungen würden wir sogar in idealistischer Anschauungsweise kaum anders auszusprechen wissen, als indem wir sagen: 'Die mit dem Charakter der Wahrnehmung auftretenden Bewußtseinsakte verlaufen so, als ob die von der realistischen Hypothese angenommene Welt der stofflichen Dinge wirklich bestände'."[3].

In quantum physics, however, we can no longer project our principles, laws and measurement results directly back into what classical physics might have called "independent reality". This is the primary implication of the so-called superposition principle and is the essence of Einstein's argument (see EPR, Bell, etc.)[4] However, the most striking quantum theoretical effect is that "the space-time description of processes, on one hand, and the classical causality law on the other hand, are complementary, mutually exclusive features of the physical processes"[5]. This is difficult to accommodate in an a priori scheme, in which both space-time describability and causality must hold in each ex-

---

[3] In our ad hoc translation: "The realistic hypothesis ... considers as existing independently on our representation that which in daily perception so appears: the material world outside us. The realistic hypothesis is undoubtedly the simplest we can build, tested and confirmed in extremely wide circles of application, sharply defined and therefore extremely useful and fruitful for practice. We hardly could express the lawlike character of our perceptions, even from an idealistic stance, except by saying: 'the acts of consciousness which appear with the character of perceptions proceed as if the world of material things presupposed by the realistic hypothesis really existed'." Helmholtz (1878), 273

[4] The superposition principle indeed states that linear combinations (superpositions) of valid states provide new, valid states for a system. This occurs also when some property (observable) has sharp but different values in the original states – from which it follows that the combined state is ambiguous concerning this observable. Were we to measure this observable we would find each time a sharp value – we don't even know what the unique event "here or there" looks like – but each trial would produce a *different* value and the final statistics of these values verify the probabilistic rules issued by the quantum mechanical analysis. It is difficult to see this result as a *one to one* mapping of some reality. This is even more pointedly expressed in the Einstein-Podolsky-Rosen argument and the discussion of the Bell inequalities. Here a situation is realized, where at a certain moment two systems together build one compound system and namely in a state which would not factorize into a product of states of the original systems (a kind of *and*: system 1 in state A *and* system 2 in state B) but would be a combination of such products (a kind of nonexclusive *or*, e.g.: system 1 in state A *and* 2 in B *or* 1 in B *and* 2 in A). Even if completely separated, neither of the two systems is in a pure state and measurements on the two (no longer causally related) systems lead to results which are correlated in a way contradicting the assumption that their properties would correspond to "elements of reality" temporarily escaping our knowledge (so-called "local hidden variables").

[5] Heisenberg (1930), 48

perience because they are the very preconditions for it (see Jammer 1988). Therefore, modern physics, while surely no longer supporting a naive realism, also does not seem to support the Kantian model – at least when understood dogmatically.

In what concerns the second point, Kant does indeed stress repeatedly the importance of the empirical element in knowledge (see, e.g., his refutation of idealism in *Kritik der reinen Vernunft*, or the discussion of the "Metaphysische Anfangsgründe der Naturwissenschaften"). But the clean separation of empirical and a priori elements in the building up of physical theories (which of course is necessary in the context of a rigid a priori structure) may be problematic. In fact we are somehow compelled to accommodate observations of the following type: Our apparently a priori space representation should be that of a 3-dimensional, euclidean space, and in 99.99% of our experiences there is no clash between this representation and the phenomenal space representation, while in 0.01% such a clash can be remarked, in the sense that spatial relations in the phenomena cannot be directly construed in terms of classical space and time (these are, of course, not the phenomena of every day experience). If the coincidence were 100% we could assume that phenomenal space is completely determined by a priori intuition (as Kant had no reasons to doubt, since no examples of the remaining 0.01% were known). If the coincidence were significantly less frequent and more random we could assume that there is an empirical element interfering with our a priori intuition of space and showing up here and there in what appears to be a clash, but may just as well be interpreted as, say, a new form of interaction. But such a good yet nevertheless imperfect agreement seems to contradict the assumed independence between a priori intuition and empirical events and raises the question, whether at least some of the peculiarities of what we consider to be our a priori intuition of space have not in fact been somehow developed in the context of a *given set* of phenomena (that is, constructed in the process of construing these phenomena), and therefore they are not a priori in the sense of Kant but imply a history and an evolution, connected with the evolution of knowledge.

To stress once more this point: That we *may* construct theories having a very different spatial structure than the one assumed to be a priori is of course no contradiction to the original Kantian model; but that we *need* to construct such theories in order to accommodate an enlarged set of observations seems to contradict this model, because the theory should always include the preconditions for experience. But, e.g., in quantum gravity or in string theories – to take an extreme, albeit still speculative example – the fundamental spacetime structure is not the usual one; the latter appears only as a derivative "phenomenon" under certain conditions. Even in the "good old fashioned" quantum theory and general relativity, which relate with sturdy – even if not daily – experience, space and time as preconditions for experience do not match our classical, daily "intuitions".

Hence the points of criticism of a dogmatic understanding of the Kantian model seem to be related somehow with its "hypostatising" human reason [Vernunft] with its a priori content.

7.   An example of a possible way to modify the Kantian model is given in Cassirer's *Philosophie der symbolischen Formen*. In comparison with the former the model offered in Cassirer's philosophy appears to acquire a more dynamic character. We shall of course not try to discuss here this approach and we shall not attempt to conclude about its success. We shall only try to see, on some points of this Ansatz, in which way it departs from the dogmatic position. It is unavoidable that these remarks will have a starkly simplified character (as the preceding too had!).

Essentially this approach appears to avoid a rigid opposition between a priori representation and empirical material by stressing that we only have to do with symbolic structures and by recognizing in all perceptual acts a "symbolic pregnance"[symbolische Prägnanz] [6]. The process of knowledge build-up appears therefore to acquire a motor, and human understanding is no longer given in its structure but becomes historical (where Cassirer speaks of culture history one may still ask whether this only means mental or includes also biological history). While the principle of causality in its most general formulation remains strongly valid, since it just represents a mental reflection about the sheer possibility of the knowledge process, other a priori representations become historical, because we can only recognize them in the invariants of the knowledge build-up process, which, however, is never closed. (Cassirer speaks

---

[6] It may be interesting to give here the full quotation, although we shall not be able to expand this point: "Unter 'symbolische Prägnanz' soll also die Art verstanden werden, in der ein Wahrnehmungserlebnis, als 'sinnliches' Erlebnis, zugleich einen bestimmten nicht-anschaulichen 'Sinn' in sich faßt und ihn zur unmittelbaren Darstellung bringt. Hier handelt es sich nicht um bloß 'perzeptive' Gegebenheiten, denen später irgendwelche 'apperzeptive' Akte aufgepfropft wären, durch die sie gedeutet, beurteilt und umgebildet würden. Vielmehr ist es die Wahrnehmung selbst, die kraft ihrer eigenen immanenten Gliederung eine Art von geistiger 'Artikulation' gewinnt – die, als in sich gefügte, auch einer bestimmten Sinnfügung angehört. In ihrer vollen Aktualität, in ihrer Ganzheit und Lebendigkeit, ist sie zugleich ein Leben 'im' Sinn. Sie wird nicht erst nachträglich in diese Sphäre aufgenommen, sondern sie erscheint gewissermaßen als in sie hineingeboren." In our ad-hoc translation: "One should understand under 'symbolic pregnance' the way in which a perceptual experience as a 'sensory' experience also incorporates a non-intuitive 'meaning' which it directly brings to expression. One is not dealing here with just some 'perceptive' data on which later on some 'apperceptive' acts will be grafted, by which the former would be interpreted, judged and reorganized. It is the perception itself which due to its inherent organization acquires a kind of 'articulation' – which, as given in itself, also belonging to a certain meaning commitment. It [the perception] is, in its full actuality, in its wholeness and vitality, also a life in 'meaning'. It will not be only later on taken over into this sphere but appears in a certain sense as born in it." Cassirer (1929), 235

of the categories as postulates of empirical thinking.) In fact Cassirer (which has a deep knowledge of contemporary physics) develops his system further to cover the question of scientific knowledge in the context of the advance of the relativity theories and was able to take over in the 1930s the discussion about quantum mechanics as a confirmation of his system (see Cassirer 1937).

8. Cassirer's system may offer a good frame for the epistemology of modern physics, although many of his concepts need further analysis. So for instance, the important concept of "symbolic pregnance" – apparently influenced by the "Gestaltpsychologie" of the turn of the century (Krois 1992) and governing especially what Cassirer calls the "expression" and "representation spheres" [Sphäre des Ausdrucks und Sphäre der Darstellung], a fundamental concept in connection with the activity of "experiencing" in the sense of Cassirer – is clearly a very difficult, complex construct, already at the level of the question of how it appears and where it is anchored. The second concept, governing the "sphere of meaning" [Sphäre der Bedeutung] and essential in the building up of scientific knowledge is that of "meaning function" [Bedeutungsfunktion][7]. The role of mathematics in the conceptual structure of physics is to be understood on the ground of this "meaning function"[8], again a very involved concept.

A wide program of research and publication activities concerning the work of Ernst Cassirer is conducted presently at FESt in collaboration with other groups (E. Rudolph, H. Wismann, A. Krois, H. G. Dosch, I. O. Stamatescu, et al.).

9. We may ask the kind of questions we started with more explicitly in connection with the development of physics. In considering the evolution of physical theories we usually speak of factual contradictions and internal inconsistencies as indicating the necessity of a new (superior) theory. Since we speak of a theory when it is secured (or at least expected to be so) in its mathematical scheme, internal inconsistency would have to do primarily with its definition as a physical theory, i.e., in relating it to the phenomena. The above distinction is therefore somewhat arbitrary. In any case, however, what in the unsatisfactory theory appears as empirical contradiction (or dependency) is

---

[7] Cassirer (1929), 332

[8] Following Duhem, Cassirer notes: "Was sich uns zunächst als eine rein faktische Mannigfaltigkeit und als eine faktische Verschiedenheit von Sinnesausdrücken darstellt, das erhält physikalischen Sinn und physikalischen Wert erst dadurch, daß wir es auf den Bereich der Zahl 'abbilden'." In our ad hoc translation: "What appears to us at first as a pure manifold of facts and as a factual diversity of sensory perceptions acquires physical meaning and physical value only when 'mapped' onto the realm of numbers." In an explicit discussion of "determinism and indeterminism in modern physics" Cassirer identifies "quantitative statements" [Maßaussagen], laws, principles and finally the "principle of causality" as the various steps of this process (see Cassirer 1937).

for the superior theory a question of internal consistency, since in this theory
the correct result must simply be *derivable*.

We do not yet understand, for instance, why there should be just four
"families" of fundamental fermions, hence, for the standard model (SM: QCD
and the unified model of electroweak interactions) this fact, as well as the
quark and leptons masses, are empirical parameters (while, e.g., the hadron
masses, which were empirical parameters for early strong interaction models, became derived quantities in SM and are hence lifted from empirical to
theoretical determination). The (yet) empirical dependence of the standard
model should again be lifted in some superior theory which would predict
with necessity the number of families, the masses of the quarks, etc.

If we abstract from the constants of nature (the light velocity c, the Planck
constant h, the Boltzmann constant k_B – which can all be set to 1 defining in
this way a consistent system of units – and in some sense also the Planck length
l_P) the number of empirical parameters should decrease in this evolution.
One speaks therefore of the Theory of Everything (ToE) which would have
no free parameter (an idea promoted by asymptotic freedom models). Now
since physics has to do with the universe it is likely that it will never end;
however, this does not mean that it is meaningless to speak of a ToE, at
least in some asymptotic sense, as a regulative idea or as an approximate,
temporary achievement.

10. The fact that superior theories have less empirical dependence than their
predecessors does not necessarily imply that if we were clever enough we would
directly find the ToE and never need empirical information. In fact the possibility of introducing measures and optimization criteria in an abstract way
in some "space of theories" seems rather limited (it has been tried, e.g., in
connection with gauge theories but this already assumed a certain frame of
"trial" theories). A more realistic view might be that of a continuous process
of gradual "transmutation" of empirical information into theoretical structure,
whereby our "affectability" [Affizierheit] itself is theoretically preconditioned
– that is, a process in which existing theoretical structures help construe experience in a way which leads to increasingly powerful theoretical structures.
Such a process appears to be in the spirit of Cassirer's view, who speaks of
the "symbolic pregnance" of the empirical facts and sees here a continuous
activity of objectification – not an accumulation of objects ("we do not recognize 'the objects', but we recognize objectively" – in our ad hoc translation
[9]).

The a priori structure as described by Cassirer is thereby not known and
actually not fixed: "die kritische Erfahrungslehre [will] die allgemeine Invari-
antentheorie der Erfahrung sein."[10] and again in our ad-hoc translation: "The

---

[9] "Wir erkennen somit nicht 'die Gegenstände', ... sondern wir erkennen gegenständ-
lich" Cassirer(1937), 286

[10] Cassirer (1910) , 356

goal of the critical analysis would be reached if we succeeded in obtaining in this way the last common part of all forms of scientific knowledge, i.e., in fixing conceptually those instants which remain stable in the progress from theory to theory because they are the preconditions for all theory."[11] The character of a "regulative idea" of this sort to pose the question of a prioricity is evident. A similar view would hold for the ToE, whose realm would be that of what Cassirer calls "terminus ad quem" – the imagined limit of the process of objectification.

11. For physical theories the conceptual structure is strongly mathematical in the sense that it must be mathematically consistent. Were this is not the case there would always be the possibility of change or even complete overthrow of the "wrong" theory. In fact one tries to secure the theory already in the process of construction, therefore one never arrives to the point of erroneously establishing a wrong theory. But it can be seen, e.g., from the present problems with quantum gravity that this program may be very difficult.

Now this can be considered as the realization of the general causality principle mentioned above, in the sense that only mathematical consistency can ensure the uniqueness and definiteness of the conceptual scheme of physical theories.

It is also interesting, however, to look for cases where the mathematical embedding may lead to discoveries by itself, apparently deprived of every foregoing "physical intuition" – such as, e.g., the prediction of antiparticles or of the Ω^-.

12. This makes mathematical structures in physics have a very special character: they cannot be purely empirical in the sense which, e.g., was opposed by Kant in the above quotation on geometry. They cannot be purely a priori either, since one should then wonder why "empirical reality" reacts so slavishly sometimes to very precise theoretical predictions based purely on the mathematical scheme (this is, extrapolated, more or less the question of why theoretical physics is possible). And it is also difficult to think of them historically, since mathematics seems to have an atemporal character (even if we do not know it all from the beginning). But if we just think of mathematical structures as tools, as tautology schemes to correlate empirical data, even if these tools were understood as very refined we would still have difficulties in understanding their role in the conceptual build-up. In fact, especially in modern physics, we must often renounce intuition in favor of mathematical description, that is, a description avoiding picturing; to quote from Richard

---

[11] "Das Ziel der kritische Analyse wäre erreicht, wenn es gelänge, auf diese Weise das letzte Gemeinsame aller möglichen Formen der wissenschaftlichen Erfahrung herauszustellen, d.h. diejenige Momente begrifflich zu fixieren, die sich im Fortschritt von Theorie zu Theorie erhalten, weil sie die Bedingungen jedweder Theorie sind." Cassirer (1910) , 357

Feynman: "How do you imagine the electric and magnetic field? ... I don't know how. I have no picture of this electromagnetic field that is in any sense accurate. ... Perhaps the only hope, you say, is to take a mathematical view. Now what is a mathematical view? From a mathematical view, there is an electric field vector and a magnetic field vector at every point in space; that is, there are six numbers associated with every point. Can you imagine six numbers associated with each point in space? That's too hard. Can you imagine even one number associated with every point? ... The whole question of imagination in science is often misunderstood by people in other disciplines. ... They overlook the fact that whatever we are allowed to imagine in science must be consistent with everything else we know: that the electric fields and the waves we talk about are not just some happy thoughts which we are free to make as we wish, but ideas which must be consistent with all the laws of physics we know." [12] So: "The most correct [way to visualize] the behavior of fields is also the most abstract: we simply consider the fields as mathematical functions of position and time."[13]

This means that the only thing we can do is to try to be aware of all possible mathematical implications involving a theoretic symbol – e.g., that of an electric or magnetic field – a kind of "nonsensual actualization" [unsinnliche Vergegenwärtigung]. This means that mathematics plays a much more direct role in the formation of the meaning structures than is suggested by the simple character of a tool. On the other hand, there is often much redundancy and arbitrariness in the mathematical scheme of a theory, due to equivalent descriptions, etc. (for instance local gauge invariance, retarded action at a distance, pilot waves, etc.). It is difficult to decide, only on the basis of the mathematical structures, which is the right choice from which one should proceed with the interpretation and the meaning giving activity. Physicists love to quote here "simplicity", "elegance", etc., but they overlook sometimes that their criteria usually mix together *physical and mathematical* aspects. All this seems to imply that the epistemological role of mathematics in physical theories is intimately related to the process of concept forming, in deep interaction with physical, ultimately empirical considerations, and no easy separation in tools and data, a priori scheme and material of experience, etc., is possible.

Let us consider again the point of view of Cassirer. It is the "meaning function" – already mentioned under point 8 above – which guarantees the intelligibility of empirical facts by projecting them on given "systems of numbers" (understood quite generally as arbitrarily complex algebraic or geometric structures). This is related, of course, with a certain understanding of "meaning" itself. For Cassirer the latter lies in the unity of mutual conceptual references under which a diversity determines itself as internal "belonging together".[14] "Meaning" is therefore an actualization of valences as they are

---

[12]Feyman (1964), 20-9.

[13]Feynmann (1964), 1-4.

[14]Since this is a central aspect it is interesting to give more details. Cassirer uncovers in his analysis the same feature of concept building showing up in all cognization

represented and made explicit by "numbers", and the "meaning of a concept"
is the decoding of the various valences which the latter acquires in the conceptual scheme. The transparency of mathematics allows the intellect [Geist]
which uses it both to be "affected" [affiziert] by the "empirical reality" and
to depart from this and free itself. The "highest" level of "sense" in concept
building is the mathematically founded distribution of valences. It is in this
way that modern physics was able to proceed and realize its concepts beyond
intuition [Anschauung]. This way of putting the problem is also clearly independent on the "subject/object" question; it simply overcomes this scheme.
It seems again that Cassirer's philosophical system can support a consistent
approach to the epistemological problems of modern physics.

13. In historical perspective, important changes in the physical theories seem
very step-like. However, even, e.g., the advent of quantum theory is not only a

---

activity: "An welchem Punkte der Erkenntnis wir die Frage auch einsetzen lassen,
ob wir bis zu ihren höchsten Stufen hinauf- oder bis in ihre untersten Schichten
hinabgehen, ob wir die Anschauung oder das reine Denken, ob wir die sprachliche
oder die logisch- mathemathische Begriffsbildung befragen: immer wieder finden
wir in ihnen jenes 'Eine im Vielen' wieder, das sich uns, als ein seinem Sinn nach
Identisches, in den verschiedensten Stufen der Konkretion darstellt und ausprägt.
Und diese übergreifend-Eine ist in all diesen Fällen nicht sowohl die Einheit der
Gattung, unter welche die Arten und Individuen subsumiert werden, als vielmehr
die Einheit der Beziehung, kraft deren ein Mannigfaltiges sich als innerlich zusammengehörig bestimmt. Diese Grundform der Beziehung ist von hervorragenden
Mathematikern als Kern des Zahlbegriffs und damit als Kern des mathematischen
Denkens überhaupt bezeichnet worden; aber sie ist in keiner Weise auf diesen Bereich beschränkt. ... sie beherrscht die Gesamtheit des Erkennens .... In ihr daher
muß der 'Begriff' begründet und verankert werden. 'Begreifen' und 'Beziehen erweisen sich der schärferen logischen und erkenntniskritischen Analyse überall als
Korrelata, als echte Wechselbegriffe." (Cassirer 1929, 348) Again in our ad hoc
translation: "At which point of the cognition process we ever put the question,
whether we climb up to its highest stairs or go down in its deepest layers, whether
we ask intuition or pure thinking, speech or logic-mathematical concept building:
we find again and again in them that 'one in many' presenting itself and making
itself obvious to us in the most different concretization steps as something identical in its understanding. And this overlapping 'one' is in all these cases not the
unity of the genus under which the species and the individuals are subsumed, but
the unity of reference by which a diversity determines itself as internal belonging
together. This basic form of reference" – but notice that Cassirer means here relations between concepts, not references to something beyond concepts – "has been
designated by outstanding mathematicians as being the nucleus of the concept
of number and hence of mathematical thinking in general; but it is by no means
restricted to this field. ... [this basic form of reference] dominates the whole of
cognition .... Therefore it is on and in it that the 'concept' has to be based and
anchored. 'Grasping' and 'referring' [in the sense of relating] appear to a sharper
logical and cognition-critical analysis always as correlates, as truly interchangeable
concepts."

question of decades itself but it also implies long pre- and post- histories. This is not a step-like process and we are warned against neglecting the gradual part of the development of conceptual structures. We should also not forget that physics generally proceeds by hypotheses building on the basis of a certain theoretical understanding and empirical information and that at least part of the conceptual development takes place in the competition between hypotheses, before one or other of them can be reasonably falsified.

Besides General Relativity, which is a classical theory of space-time and gravity, Quantum Field Theory represents the paradigm of modern physics, and the standard model of strong and electroweak interactions represents the dominant theory at present. However, this model does not unify the elementary phenomena to which it applies (does not reduce them to a unique fundamental interaction), and it leaves aside the question of quantum gravity. It is a good question whether the fundamental unifying theory may be constructed in the frame of local Quantum Field Theory, or whether it blows up this frame. But, quoting again Richard Feynman: "Or they say, 'You know those quantum mechanical amplitudes you told me about, they're so complicated and absurd, what makes you think those are right? Maybe they aren't right'. Such remarks are obvious and are perfectly clear to anybody who is working on this problem. It does not do any good to point this out. The problem is not only what might be wrong but what, precisely, might be substituted in place of it." [15] As long as we do not have a glimpse at the new physics we cannot judge upon the decisivness of the various shortcomings of the present theories.

An aspect which is very important in the development of physical theories concerns the interaction with the developments in mathematics itself. While it is difficult to follow the dynamics of this interplay, we can at least indicate the "new" stress in mathematics implied by establishing new physical theories. Thus the "new" mathematics implied by general relativity was that of general differentiable manifolds, for quantum field theory was primarily that of non-differentiability – generalized functions and operator-valued distributions. (Of course, both promoted in these contexts the use of symmetry groups, fiber bundles, operators and functional spaces, etc., in the further development of the mathematical structures of classical theories and quantum mechanics.) Following the mathematical "thinking" connected to one or other theory may be useful in understanding the conceptual changes in physics. And of course we should not forget the strong ties mathematics has to philosophy and its meaning giving role in physics arriving via this connection.[16].

It is important to notice that a very large part of theoretical physics research is concerned with proposing mathematical models and working out solutions for the fundamental equations of these models and of the general theories. This activity, which at first looks like an abstract game, is essential

---

[15] Feynman (1967), 161

[16] cf. von Weizsäcker (1992), where a farther going discussion can be found.

in two respects: On the one hand it provides the possibilities for interpreting the theory, both phenomenologically by offering predictions and a pool of proposals for adequate actualizations (e.g., the Friedmann and Robertson-Walker solutions to the Einstein equations) and theoretically by helping establish the net of mutual conceptual references inside the theoretical scheme. On the other hand it usually promotes new developments in mathematics itself, which very often at a later time have a new and sometime unexpected impact on the physics again. Characteristic also of present day research is the relevance of mathematical physics, in the frame of which very complex mathematical structures find unusual physical interest; at the same time the concern with them leads to new aspects in mathematics itself.

What is the role of mathematics in the evolution of theoretical ideas? Present day physics may be a very good field for asking this question, since with the disappointment in finding a Grand Unified Theory and even more in unifying gravity in a straightforward way there emerged a wave of models using highly nontrivial mathematical structures. There is a number of new ideas at present and some of them may turn out to lead to the right answer, that is, to the wanted fundamental theory of the known interactions. This must not necessarily be a ToE, not even in an approximate sense – although it is specific to present physics research that it involves very different domains, such as cosmology and elementary particles. Although these ideas are rather speculative, they do in fact continue and develop a point of view which already appeared in the context of the renormalization program in quantum field theory and of general relativity, namely an intimate connection between space-time, matter and interaction. The strong involvement of mathematics in present physics research brings a wide spectrum of new concepts and may be significant as indicating the search for new meaning structures. It may therefore signal rapid changes ahead.

References

Cassirer, E., 1910 , Substanzbegriff und Funktionsbegriff, Wissenschaftliche Buchgesellschaft, Darmstadt, 1990

Cassirer, E., 1929 , Philosophie der symbolischen Formen III, Wissenschaftliche Buchgesellschaft, Darmstadt, 1990

Cassirer, E., 1937 , Zur modernen Physik, Wissenschaftliche Buchgesellschaft, Darmstadt, 1987

Feynman, R. P., Leighton, R. B., Sands, M., 1964, The Feynmann Lectures on Physics, Addison-Wesley, Reading, Mass. MIT Press, Cambridge, Massachusetts

Feynman, R. P., 1964, The Character of Physical Law, MIT Press, Cambridge, Massachusetts

Heisenberg, W., 1930, Prinzipien der Quantumtheorie, B.I. Mannheim, 1958

Helmholtz, H. von, 1878 Die Tatsachen in der Wahrnehmung. In "Philosophische Vorträge und Aufsätze", Akademie-Verlag Berlin, 1971

Jammer, Max, 1988, Kants a priori and modern physics, unpublished manuscript

Kant, I., 1781, Kritik der reinen Vernunft, Wissenschaftliche Buchgesellschaft, Darm-
    stadt, 1956

Krois, J. M., 1992, Cassirer, Neo-kantianism and Metaphysics. In Revue de Meta-
    physique et de Morale, 96e an./No.4, 437

von Weizsäcker, C. F., 1992 , Zeit und Wissen, Carl Hanser Verlag, München, Wien

I

Space and Time, Cosmology and Quantum Theory

Is Conscious Awareness Consistent with Space-Time Descriptions?

Roger Penrose

Mathematical Institute, University of Oxford, Oxford, UK

1. Consciousness and Computation

It is a fact of nature that certain objects inhabiting this physical world are capable of evoking conscious awareness. Among such objects are healthy living human brains (at least when attached to human bodies). We do not know what other physical systems might be capable of this feat. There is a point of view which asserts that a sufficiently powerful computer, suitably programmed, could also evoke awareness. Indeed, it is not infrequently maintained that this conclusion would be a necessary implication of any completely scientific viewpoint. It is argued that the brain's function is simply "information processing" – the activity that computers also indulge in, sometimes very much more effectively than human brains. Although there are many activities in which the brain remains a more effective instrument than any computer system constructed to date, it is only a matter of time – so the argument runs – before computers will vastly exceed the capabilities of brains in all significant respects. Of course, opinions might differ widely as to the timescales involved. Some say that we shall be superseded in less than forty years (cf. Moravec 1988), whilst others argue that it will be many centuries, perhaps even millenia, before computers will overtake us, though eventually they must do so – if we do not destroy ourselves first!

Others (cf. Searle 1980, 1987) argue that the mere computational activity that computers indulge in would not, of itself, evoke awareness or any other mental quality. In particular, the quality of "understanding" would not be present by virtue of the mere carrying out of some calculation. Nevertheless, the possibility of a computational *simulation* of the activity of a brain while in the process of doing whatever it does when a person actually understands something is not denied by such people. But such a simulation is argued as being not *itself* capable of evoking any actual understanding or any other conscious mental quality.

The first of these viewpoints is varioulsy referred to as "functionalism" or as "strong AI" (where AI stands for "artificial intelligence"). This basic position is encapsulated in the assertion:

A. All thinking is computation, and the mere carrying out of appropriate computations will evoke feelings of conscious awareness.

The second viewpoint can be summarized as:

B. It is the brain's physical action that evokes awareness; and any physical action can be simulated computationally; but computational simulation by itself cannot evoke awareness.

It should be pointed out, however, that there is really no distinction between viewpoints *A* and *B* with regard to the question of the eventual superiority of computers over human brains. If an effective total simulation of human intelligence can be achieved – and then vastly exceeded – by some robot, then these achievements would be in no way hindered if the robot did not "actually" possess any awareness. Paradoxically, it is *A* rather than *B* that provides us with ultimate hope for a future for human awareness. If the mentality that each of us possesses is something that can be translated into some computer program, as is asserted by viewpoint *A*, then one can envisage this program being transferred to the supercomputer that controls a robot, so that the person's actual awareness would itself be transferred to robot form (Moravec 1988). Viewpoint *B* does not admit this possibility, and it leaves us with the prospect of our planet's ultimate domination by totally insentient robots.

There is, however, an important assumption involved in *B* (and also in *A*). This is that physical actions can indeed always be simulated computationally. I shall argue, in a moment, that the quality of conscious "understanding" is something that cannot even be *simulated* computationally. This goes beyond Searle's argument that no computational simulation could, of itself, evoke the internal mental quality of "understanding" since I claim that *outward* manifestations of this quality would differ from the outward manifestations of any computer-controlled robot. Thus, according to my own point of view, there must be something in the physical action of a brain, when it is in the process of consciously understanding something, that defies any computational description whatever. Accordingly, my own viewpoint asserts:

C. Appropriate physical action of the brain evokes awareness, but this physical action cannot even be properly simulated computationally.

Some might regard such a possibility as being totally outside the bounds of science. However, there are many things within precise mathematical descriptions that to defy computation. For example, there is no computational procedure for deciding whenever it is the case that a family of polynomial equations, with integer coefficients, in several integer variables, has no solution. Likewise there is no computation that can decide whenever a given set of polygonal tile shapes (say made up of squares joined together, for simplicity) is capable of tiling the entire Euclidean plane. Nor is there any computational procedure that correctly decides whenever two four-dimensional manifolds are topologically distinct from one another. It is certainly possible to produce "toy model universes" based upon such algorithmically insoluble mathematical structures (cf. Penrose 1989, p. 170). Non-computable physical laws, though unfamiliar to us, are certainly within scientific possibility. Moreover, in accordance with viewpoint $\mathcal{C}$, they might underlie whatever physical actions are taking place when a brain indulges in conscious thinking. Thus, we do not need to be driven outside the bounds of science when rejecting viewpoints $\mathcal{A}$ or $\mathcal{B}$.

In a moment I shall give what I believe are the most scientifically compelling reasons for indeed rejecting these two viewpoints. There does not appear to be any action within known physical laws that definitely requires a non-computational description, however. My own standpoint, as I shall elaborate a little at the end of this article, is that this action has to do with whatever is physically going on in the process that we unsatisfactorily describe as "the collapse of the wave function". It is my opinion that a fundamental advance in physical understanding is needed before this process can be properly described in scientific terms – and that non-computational ingredients will have to be introduced as an integral part of this advance. I must emphasize that in suggesting this I am *not* being driven to taking a non-scientific attitude, as some would maintain, such as is embodied in the alternative final viewpoint:

$\mathcal{D}$. Awareness cannot be explained in physical, computational, or any other scientific terms.

2. Non-computational Nature of Mathematical Understanding

In order to show that the conscious mind is capable of *some* kind of non-computational activity, I shall have to turn to mathematics. Once it is established that conscious *mathematical* understanding is something that cannot be simulated computationally, then my main point will have been established. It is not likely that there is anything special about mathematics in this respect. If conscious mathematical activity is seen to be something that cannot be simulated computationally, then this opens up the possibility that perhaps *all* conscious brain activity depends upon a physical action that is non-computational.

Let us consider computations in general. What is computation? It is essentially anything that can be carried out by a modern general-purpose computer. Technically this is what we call an "algorithm" or the action of a "Turing machine". Strictly speaking, such action might require an unlimited storage capacity, but I am not going to worry about this proviso here. Sometimes people consider that certain types of computational activity might come outside what is normally called "algorithmic", and they cite such things as "parallel action", "connectionism" (neural networks) "learning systems", "heuristics", "random elements" and input from the "environment". In my discussion, I am considering all these things as being effectively *included* in what I mean by computation, since they can (and often are) simulated on an ordinary general purpose computer. This certainly applies to parallel action, even for those procedures for which a parallel architecture is much more effective than a serial one. Well-defined calculational rules can also be given to simulate "learning" – such as that which underlies the action of neural networks – and also "heuristics". Random elements can be effectively simulated by using pseudo-random numbers, these being generated by perfectly clear-cut algorithms. Moreover, the environment can, in principle, also be effectively simulated – unless the relevant physics governing the behaviour of that environment is itself in some way non-computable, which is what the argument is eventually trying to establish in any case.

An important fact about computations is that some of them do not ever terminate. Consider the example:

"Find an odd number that is the sum of two even numbers".

One could clearly set a computer to perform this task. For example, it could consider all the triples (a,b,c) one after the other, ordered appropriately, where a, b and c are natural numbers (i.e. non-negative integers) for which a is odd, and b and c are even, stopping only when a triple is found for which $a = b+c$. This task would obviously never end. A considerably less obvious example of a non-terminating computation is:

"Find a number that is not the sum of four square numbers".

(By a "number", I here mean, as above, a "natural number": one of $0, 1, 2, 3, 4, 5, 6, 7, 8, 9, 10, 11, \ldots$.) The fact that this computation does not stop is far from obvious. It follows from a theorem in number theory, proved by Lagrange in the eighteenth century. If, instead, the computation had been:

"Find a number that is not the sum of three square numbers";

then the computation would have stopped at 7 (which needs four squares: $7 = 1^2 + 1^2 + 1^2 + 2^2$). For another example, we could consider:

"Find an even number greater than 2 that is not the sum of two prime numbers".

It is not known whether this computation will ever stop. The assertion that it would not is the famous Goldbach conjecture, unsettled to this day.

How do mathematicians convince themselves that certain computations in fact do not stop? In the first of our examples, we could say that this fact is "obvious", and if desired we could provide an appropriate proof; whilst in the second, the reasoning is quite involved – though it can be followed by any mathematician who follows through the details of Lagrange's proof.

When they come to their conclusions in such matters, are mathematicians always simply following some algorithm – or computation – acted out by the neurons inside their heads? I want to present an argument that renders this possibility exceedingly unlikely. It is based on the ideas of Gödel and Turing, put forward in the 1930s – basically it is a version of Gödel's famous incompleteness theorem – but only an extremely simplified distillation of these ideas will be needed here.

First, for technical reasons, we shall need to consider computations that can apply uniformly to an arbitrary natural number n, rather than having just single statements, for example:

"Find an odd number that is the sum of n even numbers";

or

"Find a number that is not the sum of n square numbers".

The first of these stops for no value of n whatever, whilst the second stops only for $n = 0, 1, 2, 3$. Now such computations can certainly be listed (computably) in some order:

$$C_0(n), C_1(n), C_2(n), C_3(n), C_4(n), C_5(n), \ldots .$$

(For example, we can take computer programs written in some standard computer language, listing them in order of length, and "alphabetically" within each specified length.) Suppose that we have some computational procedure P that, upon coming to an end, provides us with a demonstration that certain of the above computations actually do *not* terminate. The procedure P will take note of both the particular calculation C_a that it is examining and the number n that this calculation is applied to, so it depends upon two natural numbers a and n: we write this $P(a, n)$. Thus we have:

if $P(a, n)$ terminates, then $C_a(n)$ does not terminate.

It is not required that the computational procedure P be capable of *always* deciding that a non-terminating action $C_a(n)$ does not in fact terminate, but we do want to have been convinced that the procedure P does not ever give us wrong answers, i.e., that $P(a, n)$ never comes to an end when $C_a(n)$ actually does terminate. Thus, we are to think of $P(a, n)$ as some known procedure that is *sound* (and that we believe to be sound) which, when it successfully terminates, provides us with a proof that $C_a(n)$ does not terminate. We shall try to imagine that P is in fact a formalization of the totality of all the procedures available to human mathematicians for deciding that computations do not terminate, but the argument has significance *whatever* sound procedure P might happen to be. Next, consider the computation $P(n, n)$, where I have put a equal to n. This is a computation depending upon just the one variable n, so it must be in our original list; say it is the kth one:

$$P(n, n) = C_k(n).$$

Now take the particular value of n where $n = k$, so we have, from the above (with $a = n = k$):

if $P(k, k)$ terminates, then $C_k(k)$ does not terminate.

We shall try to see whether $P(k, k)$ actually terminates or not. Suppose that it does; then by the above, we see that $C_k(k)$ does not terminate. But from the displayed line before (with $n = k$), $C_k(k)$ is in fact identical with $P(k, k)$, so $P(k, k)$ does not terminate after all. This must be the right answer: $P(k, k)$ does *not* terminate, and so $C_k(k)$ does not terminate either.

The remarkable conclusion that we have drawn is that we have been able to find a calculation (namely $C_k(k)$) that *we* can see does not terminate (by the above argument), yet the calculational procedure P is unable to ascertain this fact (since $P(k, k)$ does not terminate)! Thus if P had actually been a formalization of all procedures available to human mathematicians for deciding that computations do not terminate – as we were invited to imagine above – then we should seemingly have arrived at a contradiction. We are able to see that $C_k(k)$ does not terminate whereas P fails to ascertain this fact. This conclusion depends upon our actually *knowing* the computational procedure P and, moreover, knowing that P is actually *sound*. We deduce that human mathematicians are not using a knowably sound computational procedure to ascertain mathematical truth! The mathematical insights that are in principle available to us are not things that can be reduced to computation.

There are several objections that can be (and have been) raised against this kind of argument (e.g., the many criticisms of a related discussion given by Lucas 1961; cf. Good, 1967, 1969, Benacerraf 1967, Bowie 1982, Hofstadter 1979, Lewis 1969, 1989). Some of these are misconceptions that do not really apply to the particular form of *reductio ad absurdum* that I have presented here. Others represent possible viewpoints that try to come to terms with the conclusion as I have actually presented it. Let me very briefly address these viewpoints in turn. (A much more thorough analysis will be given in a forthcoming book that I am at present engaged in writing.)[1] We might perhaps use (i) a horrendously complicated unknowable computation X, or (ii) an unsound but almost sound computation Y (which was probably Turing's own preferred solution), or (iii) an ever-changing computation, or (iv) random elements, or (v) input from the environment. On the other hand, perhaps we are indeed driven either to believing in: (vi) some mystical concept of "mind" lying outside of physical explanation (which may have been Gödel's solution) – viewpoint $\mathcal{D}$ – or else, by insisting that we remain within the realms of scientific explanations, we find ourselves driven to my own conclusion: (vii) that our (conscious) thought processes depend upon some non-algorithmic physics, a physical action that cannot even be effectively *simulated* by any computation – viewpoint $\mathcal{C}$.

---

[1] *Shadows of the Mind*, Oxford University Press, to be published

My objections to (iii), (iv) and (v) above have been outlined earlier. My difficulties in accepting (i) or (ii) stem partly from the fact that the way that we actually ascertain mathematical truth seems to be totally unlike the use of a horrendously complicated unconscious computational procedure X or of a fundamentally incorrect computational procedure Y. We must always be able in principle to break down our mathematical arguments into elementary steps that are "obvious" to our conscious understanding of the concepts involved. The discussion that I have given above shows that we cannot ever accurately and knowingly reduce such understanding to a computation. Once we have perceived how to encapsulate a body of sound mathematical insights into a computational procedure, then this very perception provides us with a *new* mathematical insight, as trustworthy as those insights that we have incorporated into our computational procedure, but which is not itself incorporated.

Let me put things another way. The most familiar form of the Gödel argument provides a statement that asserts the consistency of a set of axioms, this statement being not derivable from the axioms themselves provided that those axioms are actually consistent. If we believe that the axioms are *sound* (i.e., that the deductions from them are *true* mathematical statements), then we must believe that they are indeed consistent, so we must believe in the truth of the Gödel proposition. This insight, that enables us to deduce from soundness of a system of axioms that they must in fact be consistent, is an insight that cannot be incorporated into the axioms themselves. In general terms, it is our *understanding* of what we are doing and of what the terms of our mathematical expressions actually *mean* that enables us to break out of any formalized computational procedure. We *see* the truth of the Gödel proposition by use of our understanding of the meanings of the terms involved. People often object to using the Gödel procedure as a demonstration that mathematical insight goes beyond any computational scheme, saying that we cannot "see" the truth of the Gödel proposition unless we already know that the axioms are consistent. But the point is not that we are supposed to be able to "see" the truth of the Gödel proposition for an *arbitrary* system of axioms; it is only that we can do this for a system of axioms that we are already accepting as sound – and therefore consistent – because, by the hypothesis of the *reductio ad absurdum*, this axiom system (like the computational procedure *P*) is being hypothesized as representing the totality of the acceptable insights that are available to human mathematicians. The argument shows, quite rigorously, that one cannot form any axiom system that represents, and is believed to represent, the totality of sound mathematical insights that are available to human mathematicians.

The argument does allow, however, that we might be using an unconscious axiom system – equivalently, an unconscious algorithm – in order to derive our mathematical results. This system (algorithm) would have to be either unsound (Y above) or, if sound, its soundness would have to be unknowable to us (X above). Both possibilities strike me as being most implausible, for the process actually underlying our mathematical beliefs. In connection with Y, we should note that although mathematicians do make mistakes, these can be seen to be mistakes by the very people who make them. Once a mistake has been pointed out to a mathematician then, at least in principle, that mistake can be recognized as such. This is quite different from the blind following of an unsound algorithm. With regard to X (or Y), we should bear in mind that most of the mathematics that modern mathematicians are concerned with is enormously far from everyday experience. It is very hard to see how either X or Y could have arisen by natural selection, when our ancestors were presumably concerned with much more mundane matters (such as how to catch mammoths, etc.). How could natural selection have produced a highly sophisticated unconscious unknowable algorithm, an algorithm common to everyone in the mathematical community, that enables mathematicians to decide the truth of obscure statements which have, for the most part, almost no relevance whatever to survival? It is quite possible, however, that a general faculty for (conscious) *understanding*, be it something that could be employed towards the survival of mankind (e.g., the construction of mammoth traps) or something that ultimately was found could be used for doing mathematics (at *all* levels of sophistication), has been produced by natural selection. So long as it is accepted that the quality of understanding can be something *non-algorithmic*, then the Gödel argument is satisfied.

3. Where in Physics May We Find Non-computational Action?

I believe that these arguments indeed lend strong support to viewpoint C – assuming that we are attempting to follow the scientific path towards an understanding of the phenomenon of consciousness, and thereby rejecting D. What does this say about the physical laws that control the action of a conscious brain? It seems to me that the very least that we can conclude is that these laws must be fundamentally non-computational. By this, I do not simply mean that their consequences are unpredictable in the sense of "chaos", according to which one would need to know initial conditions to an absurdly unreasonable precision in order to predict the future behaviour of the system. As far as one can tell, chaotic systems merely introduce an effectively random element into the future behaviour; and as we have seen, randomness is not useful for taking us outside the bounds of what can be achieved by a Turing machine. After all, when one uses pseudo-random number generators, one is effectively calling upon this same feature of chaotic systems. The numbers appear to be random merely because they are "unpredictable" in very much the same sense that chaotic systems are "unpredictable".

There does remain a possibility that when the behaviour of chaotic systems is examined more thoroughly, it will be found that rather than just appearing random, they will begin to exhibit some genuinely useful non-computable behaviour, but I have seen no indications of this as yet. Much more likely, to my way of thinking, is that usefully non-computational behaviour will emerge as a necessary ingredient of an improved quantum theory, in which the unsatisfactory procedure of "wavefunction collapse" or "state-vector reduction" – which I shall denote by the letter **R** – is replaced by a mathematically sensible dynamical process. In my opinion, there is a fundamental gap in our present physical understanding between the small-scale "quantum level" (which includes the behaviour of molecules, atoms and subatomic particles) and the larger "classical level" (of macrosopic objects, such as cricket balls or baseballs). Applying at the quantum level are the procedures of unitary evolution – which I denote by the letter **U** – as can be described very precisely by Schrödinger's differential equation. But when we pass to the classical level, it seems that the world does *not* evolve according to Schrödinger's equation – a *linear* equation which preserves the superposition of "alternative worlds" that is a feature of descriptions at the quantum level. Instead, we adopt the probabilistic procedure **R**, which selects, in some appropriately random way, one alternative from amongst the multitude that are taken as co-existing in quantum-level superposition.

The issue of whether a gap in our physical understanding exists in relation to this issue is a matter of *physics* – unrelated, at least in the first instance, to the question of "minds". I believe that there is good reason, on purely physical grounds, to believe that some fundamentally new understanding is indeed needed here. But amongst physicists as a whole, this entire issue is a matter of dispute at the moment, and it would have to be admitted that the majority view would appear that *no* new theory is needed (although such outstanding figures as Einstein, Schrödinger, deBroglie, Dirac, Bohm and Bell have all expressed the need for a new theory). It is my own strong belief that a radical new theory is indeed needed, and I am suggesting that an essential component of this theory will be its fundamentally non-computational character (cf. Penrose 1989).

An important feature of any "realistic" modification of quantum theory, in which the procedure **R** is replaced by some well-defined physical action, is that this action must be *non-local* in a way that is consistent with the type of violation of Bell's inequality that has been observed in actual experiments (cf. Aspect 1976, Aspect and Grangier 1983). In these particular experiments, pairs of photons are emitted in opposite directions from a source (where the source consists of atoms that undergo transitions from one energy level to another), the photons being individually detected in two separate places, some twelve meters apart, at opposite sides of the source. The detectors are arranged to measure the components of the photons' polarization at certain orientations, where the precise choice of orientation is not made until the photons are in full flight from source to detectors. It is found (in accordance with Bell's 1964 analysis of such situations) that the statistical correlations between the polarization measurements at the two detectors is inconsistent with the pairs of photons being separate independent entities, but each pair of photons consists of a single "connected" entity, at least up until the choice polarization measurement is made on one of the photons or the other. This is despite the fact that the two photons are many meters apart when this choice is indeed made.

The standard quantum-mechanical "picture" is of a state vector that describes each photon *pair* as evolving according to $\mathbf{U}$ – the deterministic computable Schrödinger equation – up until the moment that a measurement is first made on a member of the pair. At that moment one evokes the *second* quantum-mechanical procedure $\mathbf{R}$ (statevector reduction), which is no longer deterministic, but *probabilistic*, according to standard theory. A measurement of one photon's polarization will immediately force the polarization of the *other* photon into a particular state, which is the one that is then measured by *that* photon's detector. Somehow, it would seem that a "message" is sent instantaneously from one photon to the other, signalling how that other photon is to arrange its polarization. This "message" does not transmit actual information concerning the choice of polarization that is made in the first measurement – in the sense that one cannot use this effect to send an instantaneous signal from one place to another – but without postulating some kind of "message" of this kind, one cannot make sense of the statistics that relate the polarization measurements on the two photons. Instead of thinking in terms of "messages", one can think of the two photons as being, in a sense, one "connected entity", but this really amounts to a restatement of the same thing.

An essential difficulty with such non-locality arises if we try to make a relativistically invariant picture of what is happening. With superluminary "messages" of this kind, it is not clear which photon is sending the "message" and which is receiving it. In the standard quantum-mechanical state-vector description, it is the first photon measured that determines the *time* at which the $\mathbf{R}$ procedure is enacted, so the other photon encounters this "reduction" phenomenon *before* it is actually measured. But in another Lorentz frame, things can be the other way about, and it is now the photon that had previously been called the "first" photon that encounters $\mathbf{R}$ before it is actually measured. It should be made clear, that the actual probabilities come out the same whichever Lorentz frame is used, so there is no physical *contradiction* with relativity here. Nevertheless, the situation poses a profound challenge to any "realistic" space-time picture of what is going on, consistently with the tenets of relativity theory. Relativity tells us that physical effects are supposed not to propagate faster than light signals. The mysterious quantum *entanglement* that exists between the two photons is something that indeed defies a normal relativistically invariant space-time description – though various "abnormal" descriptions are possible. The simplest such "abnormal" description that I have come across involves *two* state vectors to describe any system, one of which evolves forward in time, from a measurement, in the normal way whilst the other evolves *backwards* in time. (This was described to me by Yakir

Aharonov in 1991). Other possibilities that I have encountered seem no less strange.

Whatever the merits of such proposals, it is my own strong opinion that when a sensibly "realistic" picture of what is involved in the quantum-mechanical **R** procedure finally comes to light, it will be found that not only does this involve non-computable action of some kind, but that the description will indeed turn out to be at variance with our presently conventional space-time view.

4. Time and Conscious Perceptions

Returning to the issue of our conscious perceptions, it should be recognized that there is something very odd about our perceived "flow of time", from the point of view of modern physical theory. According to modern physics, "time" is merely a particular coordinate measure in the description of the location of a space-time event. There is nothing in the physicist's space-time descriptions that singles out "time" as something that "flows". Indeed, physicists quite often consider model space-times in which there is only *one* space dimension in addition to the single time dimension; and in such two-dimensional space-times there is nothing to say which is space and which is time. Yet, no-one would consider space to "flow"! It is true that one often considers time-evolutions in physical problems, were one may be concerned with computing the future from the present state of the system. But this is not at all a necessary procedure, and calculations are normally carried out this way *because* one is concerned with modelling, mathematically, our experiences of the world in terms of the "flowing" time that we seem to experience. It is our apparent experiences that tempt us to bias our computational models of the world in terms of time-evolutions – frequently, but not invariably – whilst the physical laws themselves do not contain such a compelling inbuilt bias.

Indeed, one could say that is only the phenomenon of *consciousness* that requires us to think in terms of a "flowing" time at all. According to relativity, one has just a "static" four-dimensional space-time (admittedly with some differential equations relating behaviour at different space-time events), with no "flowing" about it. So consciousness appears to have some special role in relation to the passage of time.

On the other hand, it would be unwise to make too strong an identification between the phenomenon of conscious awareness, with its seeming "flowing" of time, and the physicist's use of a real-number parameter t to denote what would be referred to as a "time coordinate". In the first place, it is clear that the precise concept of a "real number" is not completely relevant to our conscious perception of the passage of time, if only for the reason that we have no sensibility of very tiny time-scales – say time-scales of even just one hundredth of a second, for example – whereas the physicist's time-scales hold good down to some 10^{-25} seconds, or perhaps down to 10^{-53} seconds, and the mathematician's concept of time as a *real number* would require that there be *no* limit of smallness whatsoever, below which the concept meaningfully applies.

Quite apart from these issues, there are some puzzling experimental findings about the *timing* of conscious experiences that have almost paradoxical implications. In the mid-1970s, H.H. Kornhuber and his associates (cf. Deeke *et al.* 1976) used electroencephalograms (EEGs) to record electrical signals at a number of points on the heads of several human volunteers, who were asked to flex the index finger of their right hands suddenly at various times *entirely of their own choosing*. The idea was that the EEG recordings would indicate something of the mental activity that is taking place within the skull, and which is involved in the actual conscious decision to flex the finger. In order to obtain a significant signal from the EEG traces, it was necessary to average the traces from several different runs, and the resulting signal was not very specific. However, what was found was that there was a gradual buildup of recorded electric potential for a *full second*, or perhaps even a second and one-half, *before* the finger was actually flexed. This seemed to indicate that the conscious decision process takes a second or more in order to act. This is to be contrasted with the much shorter time that it takes to respond to an external signal if the mode of response has been laid down beforehand. For example, instead of it being "freely willed", the finger flexing might be in response to the flash of a light signal. In that case a reaction time of about one-fifth of a second would be normal – about five times faster than the "willed" action that is relevant to Kornhuber's data. Thus, according to these experiments, and also more recent experiments of a similar nature performed by Libet and others, it would appear that conscious "willed action" takes about a second in order to act.

In a second experiment, Libet (1979), in collaboration with Feinstein, tested subjects who had had to have brain surgery (for some reason unconnected with the experiment) and had consented to having electrodes placed in parts of the brain concerned with receiving sensory signals. The upshot of Libet's experiment was that when a stimulus was applied to the skin of these patients, it took about half a second before they were consciously aware of that stimulus, despite the fact that the brain itself would have received the signal of that the stimulus in only about a hundredth of a second, and a pre-programmed "reflex" response to such a stimulus (cf. above) could be achieved by the brain in about a tenth of a second. Moreover, despite the delay of half a second before the stimulus reaches awareness, there would be the subjective impression by the patients themselves that they had already been aware of the stimulus at an earlier time – essentially simultaneously with the moment that it had been applied.

Each of these experiments, by itself, would not be paradoxical, although maybe a little disturbing. Perhaps one's conscious decisions to act *do* need to occur something like a whole second before that action finally takes place; perhaps one's passive consciousness *does* require something like half a second before it can be evoked. But if we take these two findings together (and take each of them at its face value) then we seem to be driven to the conclusion that in any action in which an external stimulus leads to a consciously controlled response, something like a second and one half, at least, is needed before that response can occur.

This assumes that there is an "actual time" at which a conscious experience takes place, and that time must precede the "actual time" at which the conscious response to that experience occurs. This indeed would have seemed to have been our natural expectation. However, the Kornhuber-Libet experiments appear to lead us into some difficulties with this "natural expectation". For there are many instances of human activity that have the appearance of being consciously controlled, yet it takes far less than a second and one-half between the initial stimulus and the resulting response. Perhaps the most commonplace of these is in ordinary conversation, where consciously controlled response seems to take place in a fairly small fraction of a second.

No doubt there are many plausible loopholes in this apparent paradox, and perhaps nothing mysterious about the timing of consciousness need be inferred. My own suspicion, however, is that there could indeed be something unexpected and important underlying these experimental findings. Consciousness is already seen to have an odd and unique relationship with the (apparent?) passage of time. I believe that there may well be something else that is odd about this relationship, and it may not even be meaningful really to attach a "time" to a conscious experience. If it is the case, as I have being trying to argue above, that the phenomenon of consciousness is dependent upon some physical process that underlies the **R** procedure of quantum mechanics, and bearing in mind that the non-locality and non-causal aspects of this procedure are inherent in quantum-entangled systems, we should indeed be led to expect something odd in the relationship between consciousness and time.

I had better leave my speculations at that, for the moment. Perhaps further developments in theory and in physical and neurophysiological experiments will shed more light on these questions in the future.

References

Aspect, A. (1976): Proposed experiment to test the nonseparability of quantum mechanics, *Phys. Rev.* **D14**, 1944-51.

Aspect, A. and Grangier, P. (1986): Experiments on Einstein-Podolsky-Rosen-type correlations with pairs of visible photons, in *Quantum Concepts in Space and Time*, eds. R. Penrose and C.J. Isham (Oxford University Press, Oxford).

Bell, J.S. (1964): On the Einstein-Podolsky-Rosen paradox, *Physics*, **1**, 195-200.

Benacerraf, P. (1967: God, the Devil and Gödel, *The Monist* **51**, 9-32.

Bowie, G.L. (1982): Lucas' number is finally up, *J. of Philosophical Logic* **11**, 279-285.

Deecke, L. Grötzinger, B., and Kornhuber, H.H (1976): Voluntary finger movement in man, cerebral potentials and theory, Biolog. Cybernetics **23**, 99.

Good, I.J. (1967): Human and machine logic, *Brit. J. Philos. Sci.* **18**, 144-147.

Good, I.J. (1969): Gödels's theorem is a red herring, *Brit. J. Philos. Sci.* **18**, 359-373.

Hofstadter, D.R. (1979): *Gödel, Escher, Bach: an Eternal Golden Braid* (Harvester Press, Stanford Terrace, Hassocks, Sussex).

Lewis, D. (1969): Lucas against mechanism, *Philosophy*, **44**, 231-3.

Lewis, D. (1989): Lucas against mechanism II, *Can. J. Philos.*, **9**, 373-6.

Libet, B. (1979): Subjective referral of the timing for a conscious sensory experience, *Brain* **102**, 193.

Lucas, J.R. (1961): Minds, Machines and Gödel, *Philosophy* **36**, 120-124; reprinted in Alan Ross Anderson (1964): *Minds and Machines* (Prentice Hall, Englewood Cliffs).

Moravec, H. (1988): *Mind Children: The Future of Robot and Human Intelligence* (Harvard University Press, Cambridge, Mass.; London).

Penrose, R. (1989): *The Emperor's New Mind* (Oxford University Press, New York, Oxford).

Searle, J.R. (1980): Minds, Brains and Programs, in *The Behavioral and Brain Sciences, Vol. 3* (Cambridge University Press, Cambridge).

Searle, J.R. (1980): Minds and Brains Without Programs, in *Mindwaves*, eds., C. Blakemore and S. Greenfield (Basil Blackwell, Oxford).

Space and Time: A Privileged Ground for Misunderstandings Between Physics and Philosophy

François Lurçat

Laboratoire de Physique Théorique et Hautes Energies Université
Paris XI, bâtiment 211, 91405 Orsay Cedex, France

As I understand, one of the aims of this workshop is to strive for a better understanding between physicists and philosophers. My contribution will be to describe some misunderstandings between physics and philosophy, or else inside physics on philosophical grounds, in the field of the workshop.

A surprising field indeed. It has witnessed, in our century, several major breakthroughs, both in physics and philosophy. There are, I believe, deep convergences between what was achieved by Einstein and Bohr on the one hand, and by Bergson and Husserl on the other hand; but they have been, as far as I know, virtual convergences: the efforts went in the same direction, but they did not meet. In some cases the proximity between physical and philosophical achievements went unnoticed; in some other ones, it has been hidden by inessential errors, or by disagreements about other matters.

The Einstein-Bergson Discussion

1. In 1922, Bergson published a book, *Duration and Simultaneity*, in which he criticized the relativistic notion of multiple times. There is, he contended, only one time in the universe; the plurality of times implied by the special theory of relativity is merely an "effect of perspective". This claim was of course strongly criticized by several physicists.

Today we may take a short cut across these discussions, because we have many experimental proofs of the multiplicity of times. The most direct one is perhaps the experiment made in October 1971 (Hafele and Keating, 1972), in which four cesium beam atomic clocks were flown on jet flights around the earth. The multiplicity of times is established here beyond doubt; the time difference is of the order of 50 nanoseconds, for 40 or 50 hours of flight. So

Bergson was wrong: in suitable conditions, the plurality of times is more than a mere effect of perspective[1].

2. Is that all? I do not think so. I think that Bergson's point of view brings something new and relevant to our understanding of the nature of time. For short, I shall use the discussion (or more exactly the exchange) between Einstein and Bergson, which took place at the Société Française de Philosophie on April 6, 1922 (Bergson, Einstein et al., 1922). (I use the translation of Gunter, 1969).

"Common sense, Bergson said, believes in a single time, the same for all beings and all things. What does such a belief stem from? Each of us feels himself endure: this duration is the flowing, continuous and indivisible, of our inner life. But our inner life includes perceptions, and these perceptions seem to us to involve at the same time ourselves and the things. We thus extend our duration to our immediate material surroundings. Since, moreover, these surroundings are themselves surrounded, there is no reason, we think, why our duration is not just as well the duration of all things. (...) The idea of universal time, common to minds and to things, is a simple hypothesis. But it is a hypothesis that I believe to be well founded and which, in my opinion, contains nothing incompatible with the theory of relativity".

Now there are two possible definitions of simultaneity. Suppose I perceive two flashes departing from two points. "I term them simultaneous because they are *one* and *two* at once: *one*, in so far as my act of attention is indivisible; *two*, in so far as my attention nevertheless divides itself between them and doubles without splitting itself". This is the intuitive definition. The relativistic definition is different: "Two events more or less distant (...) are here called simultaneous (...) when they correspond to an identical indication, given by two clocks which are found next to each of them". These clocks have been synchronized by the well known procedure.

Bergson now asks: "In posing this second definition of simultaneity, (...) does not one admit the first implicitly alongside of the second?". Let E and E' be the two events, H and H' be the clocks placed respectively next to each of them. Einstein's definition of simultaneity works for the indications of H and H'. But the simultaneity between E and the indication of H, for instance, is given by a perception which unites them in an indivisible act, according to the first definition. "If this simultaneity did not exist, the clocks would count for nothing. Clocks would not be made, or at least no one would buy them. For clocks are only bought in order to know what time it is; and "to know what time it is" consists in observing a correspondence (...) between an indication of a clock and the moment at which one finds oneself, the event taking place – something, finally, which is not the indication of a clock".

Finally, Bergson stresses that he has no objection against Einstein's definition of simultaneity, neither against relativity theory in general. But "once

---

[1] Other verifications have been made, using for instance muon decay (Bailey et al., 1977), or a hydrogen maser (Vessot et al., 1980).

relativity theory is accepted as a theory in physics, everything is not finished. It remains to establish the philosophical signification of the concepts it introduces. It remains to discover at what point the theory renounces intuition, up to what point the theory remains attached to it. (...) It will be seen, I believe, that relativity theory contains nothing incompatible with the ideas of common sense".

3. Einstein's answer was much shorter. "The question," he said, "is therefore posed as follows: is the time of the philosopher the same as that of the physicist? The time of the philosopher is both physical and psychological at once; now, physical time can be derived from the time of consciousness. Originally individuals have the notion of the simultaneity of perception; (...) but there are objective events independent of individuals, and, from the simultaneity of perceptions one passes to that of events themselves. In fact, that simultaneity led for a long time to no contradiction due to the high (...) velocity of light. The concept of simultaneity therefore passed from perceptions to objects. To deduce a temporal order in events from this is but a short step, and instinct accomplished it. But nothing in our minds permits us to conclude to the simultaneity of events, for the latter are only mental constructions, logical beings. Hence there is no philosophers' time; there is only a psychological time different from the time of the physicists".

4. A few critical comments are now in order. Let us first pay tribute to Bergson: his remark about the relation between both definitions of simultaneity agrees with Einstein's original paper (Einstein, 1905) on special relativity. We read there: "If at the point A of space there is a clock, an observer at A can determine the time values of events in the immediate proximity of A by finding the positions of the hands which are simultaneous with these events. (...) But it is not possible without further assumption to compare, in respect of time, an event at A with an event at B. We have so far defined only an "A time" and a "B time". We have not defined a common "time" for A and B (...).

That is, for events which take place in the "immediate proximity" of a point A or B, the relativistic definition of simultaneity is irrelevant: here it is the observer who manages to notice simultaneities, and he does it according to the usual life procedures, correctly analysed by Bergson.

But is not Einstein's recourse to an observer purely pedagogical? Cannot we do without any observer, using only suitable instruments to record simultaneities without any human intervention? – Yes, we can, at least from a strictly logical point of view. We can do without observers, and without the usual definition of simultaneity, stressed by Bergson. (This was pointed out especially by André Metz, one of Bergson's opponents; see Gunter, 1969, pp. 187–188). But if we are concerned, as were both Einstein and Bergson, not only with logic but also with meaning, if we want to understand what we are talking or experimenting about, if we want to be able to teach it – then we must necessarily use the usual, intuitive notion of simultaneity as a starting point. Let us recall that the requirement that mathematical description of motion should have a meaning plays an essential role in Einstein's 1905 article. (See Einstein, 1905, § 1). The same requirement was in Bergson's mind when he ended his speech by the statement that once relativity theory is accepted as a theory in physics, everything is not finished, etc. (see above, § 3).

5. What is the origin of Bergson's error? There are probably several relevant answers to this question; let me give at least one. Bergson rightly puts forward the subjective side of time, the internal duration, the flow of consciousness. But the subjective aspect of time cannot be reduced to internal duration without any external reference, because we are not pure consciousnesses, and our lives are not entirely devoted to introspection. Subjective time is shaped by the day-night rhythm and by the succession of the seasons. Forgetting this side of subjective time is a very strange position indeed. Perhaps it may be traced back to Augustine, who in the book XI of the *Confessions* discards the astronomical definition of time because of Joshua: during the battle of Gabaon the sun was still, but the time flowed, since the battle was brought to an end in due time (XI, 23). The only solution Augustine sees is then a purely subjective and individual one: in thee, my soul, I measure time (XI, 27). Perhaps we have here one of the first damages due to a literalist exegesis of the book of Joshua, a reading which has also caused much trouble in the time of Galileo.

Now what does all this have to do with Bergson? One of his main points is the existence of a universal time, common to all consciousnesses. He tried to make relativistic physics compatible with this thesis, and failed. Furthermore, he considered the use of reference systems in physical theory as fictitious, albeit necessary. He disliked solids (as far as their role in thought is concerned): he held their rigidity to be contrary to the very nature of life. It was therefore difficult for him to realize that reference systems, materialized by rigid solids, are indeed a basic element of modern physical theories. More specifically, he could not see that there is really a time which, for us human beings, is universal; it is the time of the reference system defined by the earth. This time is the time of common sense, the time of the *Lebenswelt*, the everyday world; the time of *Works and Days*, as Hesiod put it. The time of physicists and astronomers (which is now the International Atomic Time) is an extreme quantitative refinement of this time of common sense.

Bergson did not realize that relativity theory disclosed an intimate relation, ignored by prerelativistic physics, between the earth and our time. It is even more surprising that, as far as I know, nobody among his opponents in the dispute about Einstein pointed out this to him. The connection between time and earth is perhaps one of the most interesting achievements of relativity theory. Its neglect or underestimation may be due to an absolutisation of Copernicanism, which will be discussed later. Today it is reinforced by the influence of some fashionable cosmological speculations.

Husserl and Space

6. In May 1934 Husserl wrote a manuscript which he never published. The following comment was written on the envelope: "*Overthrow of the Coperni-can doctrine* in its usual interpretation as a world view. The primitive arch earth does not move. Basic researches about the phenomenological *origin of the corporeality and spatiality of nature* in the first scientific sense. All necessary founding researches"[2]. The original text was published in 1940 by Marvin Farber (Husserl, 1940). A French translation was published in 1984, then in book form in 1989 (Husserl 1984, 1989). I do not know whether there is an English translation. Here are some typical excerpts (I apologize for my uncertain translation).

"We Copernicans, men of the modern times, say: the earth (...) is one of the heavenly bodies of the infinite world space. The earth is a body of spherical form which, of course, is not perceptible at a single time and by one person. (...) However it is a body! But for us it is the ground of experiencing all bodies in the experienced genesis of our representation of the world. This "ground" is not at first experienced as a body, it becomes ground-body at a higher level of the constitution of the world from experience, and this overtakes its original ground form".

"(...) The motion takes place on the earth itself, it starts from it. The earth itself, in the original form of the representation, is not in motion and not at rest, it is with respect to it that rest and motion have first a meaning. It is only afterwards that the earth "moves" or rests, and the same holds true in respect of the heavenly bodies and the earth as one of them".

"(...) But if it is so, can we say with Galileo: *eppur si muove* ? And not, on the contrary, that it does not move? Of course it is not at rest in space while having the possibility of moving, but, as we tried to show above, it is the arch which first makes possible the meaning of any motion and of any rest as a mode of motion. However its rest is not a mode of motion"[3].

---

[2] "*Umsturz der kopernikanischen Lehre* in der gewöhnlichen weltanschaulichen Interpretation. Die Ur-Arche Erde bewegt sich nicht. Grundlegende Untersuchungen zum phänomenologischen *Ursprung der Körperlichkeit der Räumlichkeit der Natur* im ersten naturwissenschaftlichen Sinne. Alle notwendige Anfangsuntersuchungen."

[3] "Wir Kopernikaner, wir Menschen der Neuzeit sagen: Die Erde (...) ist einer der Sterne im unendlichen Weltraum. Die Erde ist ein kugelförmiger Körper, freilich nicht auf einmal und von Einem wahrnehmbar (...) Doch ein Körper! Obschon für uns der Erfahrungsboden für alle Körper in der Erfahrungsgenesis unserer Weltvorstellung. Dieser "Boden" wird zunächst nicht als Körper erfahren, in höherer Stufe der Konstitution der Welt aus Erfahrung wird er zum Boden-Körper, und das hebt seine ursprüngliche Boden-form auf." "(...) Auf der Erde, oder an der Erde, von ihr weg, auf sie hin findet Bewegung statt. Erde selbst in der ursprünglichen Vorstellungsgestalt bewegt sich nicht und ruht nicht, in bezug auf sie haben Ruhe und Bewegung erst Sinn. Nachher aber "bewegt" sich oder ruht

7. This text is not easy to understand, especially for physicists. I will try to give some philosophical comments about it; but first, I would like to quote an excerpt of an interview of a psychologist (Lurçat (L.), 1984, p. 147) with a girl, Gaëlle, aged four years and nine months. The subject of the discussion was the animated cartoons about "space", space travels and so on, which can be seen on TV. It should be noted that for children of that age, these cartoons are the only referent of the word "space".

"*Are we in space ?* – No, we are not in space, we are in life. Life is not space, life is on the earth. The earth exists, the moving earth does not exist. Because in the Goldorak picture the earth moves, but that is not true, it exists in Goldorak, it exists in the Village in the Clouds, but it does not exist here".[4]

Half a century after Husserl, notwithstanding so many technical and scientific advances, Gaëlle takes essentially the same position as Husserl. It remains to be seen whether this is because she is a young child, and because Husserl refused to take into account the achievements of science since Copernicus. Or whether, on the contrary, Husserl has pointed out something which is more fundamental than science, and which allows to give a meaning to the theories of physics and some other sciences.

8. Does the earth really move? The theoretical and social status of this question is rather strange nowadays. Sociologists and such people make polls, which prove, they say, that "most of the public appear not to have caught up with Nicholas Copernicus and Galileo Galilei" (Theocharis, 1989). As Theocharis points out, this might also prove that sociologists have not caught up with Einstein. One finds indeed in the popular book by Einstein and Infeld the statement that according to general relativity theory, the sentences "the sun is at rest and the earth moves" and "the sun moves and the earth is at rest" simply mean two different conventions concerning two coordinate systems. But the question is not so simple after all. Perhaps the equivalence stated by Einstein and Infeld is valid at the level of general principles (and perhaps not: see, for instance, Fock (1947) and Ginzburg (1974)). But if we have to study a concrete problem connected with the relative motion of the earth and the stars, we shall certainly use the heliocentric system, not the geocentric one. (How can one describe the aberration or the parallax of the stars in the geo-

---

Erde – und ganz ebenso die Gestirne, und die Erde als eines unter ihnen." "(...) Aber wenn dem so ist, dürfen wir mit Galilei sagen, daß *eppur si muove*? Und nicht im Gegenteil, sie bewegt sich nicht? Freilich nicht so, daß sie im Raume ruht, obschon sie sich bewegen könnte, sondern wie wir es oben darzustellen versuchten: sie ist die Arche, die erst den Sinn aller Bewegung ermöglicht und aller Ruhe als Modus einer Bewegung. Ihr Ruhen aber ist kein Modus einer Bewegung."

[4] "*Nous, on est dans l'espace*? – Non, on n'est pas dans l'espace, on est dans la vie. La vie c'est pas l'espace, c'est par terre la vie. La terre ça existe, la terre qui bouge ça existe pas. Parce que dans le film de Goldorak la terre elle bouge, mais c'est pas vrai ça, ça existe dans Goldorak, ça existe dans le Village dans les Nuages, mais ça existe pas ici."

centric system? It must be very intricate). Notwithstanding general relativity, the two systems are not *really* equivalent.

9. But however interesting this question may be, it is not my main point here. Husserl did not mean to overthrow Copernicanism as a scientific theory, but as a world view ("in der gewöhnlichen weltanschaulichen Interpretation"). His point was that the world view of Gaëlle did not lose its legitimacy because of the attainments of Copernicus, Galileo, Newton or Einstein. This does not mean that scientific theories are illegitimate. It means that they are not able to found themselves. That is, in Husserl's terms: "The thinking pertaining to physics establishes itself on the foundation laid by natural experiencing (or by natural positings which it effects)" (Husserl, 1982, § 52, p. 100 of the original text). This should not be perceived by physicists as a manifestation of philosophical *hubris*, or as a claim for priority. We, physicists, are not only physicists, but also citizens, and often professors; we are not only concerned with the development of physics, but also with its practical and intellectual use. We are, or we should be, concerned with the fact that physics is less and less understood. Husserl's theoretical position meets the practical requirements of pedagogy. Anybody who happened to teach physics or mathematics will perhaps agree that if we want students really to *understand* what they are taught, we have to lean on their "natural experiencing" instead of merely giving them the results of scientific experimentation and speculation.

Similarly, if we want to understand (and help to solve) the problems posed by the present development of sciences, we should not remain entirely inside science, or more exactly: we should not remain entirely inside the mental habits of the scientific community. This is one of the reasons why Husserl's philosophy is highly relevant for us today. It is a way of thinking which is both outside and inside science. It has grown out of a meditation about mathematics, logic, and physics. Husserl was at first a mathematician; his teacher was Weierstrass who completed the rigorization of analysis, a major event in the history of mathematics. On the other hand, he soon began to criticize some prejudices of the scientific community of his time. His struggle against psychologism implied a criticism of the so-called scientific psychology. (This criticism has been further developed by several psychologists and psychiatrists, especially in Germany and in the Netherlands; see, for instance, the beautiful book of Erwin Straus, 1956, 1989; also Kockelmans, 1987). His work on logical problems implied a criticism of the realism of mathematicians. And his whole work implied a criticism both of the realism and idealism of physicists.

10. Now back to our manuscript, to our *Umsturz* (overthrow) of Copernicanism as a world view. What did physicists say in answer to this bold attempt? As far as I know (and I would like to be wrong), they said just nothing. Here the misunderstanding reaches its highest level: there is no understanding at all, but mere deafness.

Let us try, however, to understand how Husserl's approach to the motion of the earth (a paradox for physicists, a return to common sense for the layman) is connected to the main features of his philosophy. As we have seen, phenomenology implies a reversal of the usual conception of the relation between science and everyday life: the latter should not be discarded because it is related to prescientific conceptions; we should, on the contrary, realize that it gives the proper foundation for the "thinking pertaining to physics". Now phenomenology implies a second reversal, even more important. It requires us to reverse the classical conception of the relation between appearance and essence.

The classical attitude was to consider appearances as Sherlock Holmes considered cigarette ends: merely as clues. The aim of knowledge was to go beyond appearances: they stood in the way of true knowledge, they had to be discarded or overcome, just as when we receive a letter, we begin by tearing the envelope open and throwing it away – it has nothing to teach us about the content of the letter.

For Husserl, on the other hand, the classical opposition between essence and appearance does not hold any more, or at least it must be completely reconsidered. The appearances themselves turn out to be essential: in Husserl's words, they are "eidetic necessities" (Husserl, 1982, § 150, p. 315 of the original text). They must be studied for themselves, in their multiplicity and variability. Husserl has given special attention to the case of interest for us here: that of a physical thing. "Something such as a physical thing in space, he writes, is only given in multiple but determined changing "perspective modes" and, accordingly, in changing "orientations" not just for human beings but also for God – as the ideal representative of absolute cognition" (Husserl, loc. cit.).

11. As far as space is concerned, the phenomenological approach first implies that it should not be identified with a mathematical space (euclidean or other) given *a priori*. Space is given by a construction, and the way it is built says us what it is.

Now how do we acquire space? We do not start with space by itself, but with the spatial properties of things, their relations between themselves and with us as embodied persons.

Furthermore, we do not acquire the spatial properties of things in a purely visual way, but by the intimate combination of visual and kinesthetic exercise. (Recall that kinesthesia is the perception of the configurations and motions of our own body). This had been understood already by the Scottish philosopher Thomas Reid (1710–1796), who had imagined plants endowed with human intelligence; the only information they get about the world is visual. The natural geometry of these *Idomeneans* would be spherical, and not Euclidean. (After Imre Toth, 1977; see also N. Daniels, 1972).

Already in his 1907 lectures *Thing and Space*, Husserl explains that any spatiality is constituted only in motion – the motion of the object itself and the motion of the "I". Of course in the individual, "elementary" perception, the thing appears in its spatial body and in its place with respect to other things. But here it is not yet given fully, in the evidence of its identity. The logical synthesis which *identifies* the object can be accomplished only on the ground of the continuous synthesis of the multiple perceptions, and this, as we have said, requires motion (Husserl, 1973, pp. 154–155).

In Husserl's later work, especially in *Ideen I* (1913), the multiple perceptions received the name of *Abschattungen* (adumbrations). The lawful variation of the adumbrations, regulated in function of the motion of the body, of which kinesthesia gives an account, is called the concordance of the adumbrations. The object is given as a result of the synthesis carried out by the intentional activity of the consciousness. This way of being given by adumbrations characterizes the essence of the spatial thing (a lived experience, for instance, is not given by adumbrations). (Husserl, 1982, pp. 77–78 of the original text).

Here it would be interesting to compare Husserl's conception of space to that of psychologists like F.J.J. Buytendijk (who joined phenomenology), or H. Wallon (who ignored it). I lack time and competence to do this. Let me only point out that Buytendijk stresses the difference between physical space and vital space (Buytendijk, 1957). As to Wallon, he considers usual space, not as a primitive notion, but as a construction, realized by the interplay of different sensitivities. The elementary intersensorial and interpostural syntheses begin with the first gestures of the child; their development takes place during the first two years of his life. The real space which lies before us is already the product of these primary combinations (Wallon, Lurçat, 1987, p. 109).

12.   After this much too short excursion through the Husserlian study of space, we may give a new glance to the 1934 manuscript. Clearly, Husserl did not side with the Holy Office against Galileo. He had undertaken to establish "a universal philosophy, which can supply an organum for the methodical revision of all the sciences" (Husserl, 1947). One of his aims was to find again a meaning for the theoretical constructions of the physical sciences. The problem of meaning was central in his thought, as it is today for us (not only as physicists or philosophers, but also as citizens and as human beings). He criticized the evacuation of meaning (*Sinnentleerung*) in physics, due to its technicization (Husserl 1954, § 9g). The process of technicization has gone faster and faster since Husserl's death. Suffice it to quote a single example: we have quantum mechanics, a theory of microscopic phenomena which is wonderfully powerful and versatile; but, as Feynman put it in a much quoted sentence, "nobody understands quantum mechanics". Now if really nobody understands it, are we sure that it makes sense? And does it deserve to be called a science? Husserl's answer to the latter question in definitely "no": for him, a science with paradoxes or with unclarified basic concepts it is not a science, but a theoretical technique (Husserl, 1974, p. 161).

The attempt to overthrow Copernicanism as a world view is a particular instance of the struggle for meaning. In his study *The origin of geometry*

(1954, Appendix III), Husserl puts forward as an aim the reactivation of the original activities enclosed in the founding concepts. If we apply this idea to space, it will mean going back to the syntheses which allow us to build our lived space. These syntheses take place on the ground, because (as Gaëlle put it) "life is on the earth". Thus Husserl, when he sides with common sense against a misunderstood Copernicanism, strives to preserve both the intuitive conditions of everyday life and the conditions of a return of physics towards meaning.

Quantum Space: Vacuum and Its Uncertain Status

13. What did quantum mechanics bring new about space and time? Let is first recall some pre-quantum ideas about ether. In his 1905 paper about relativity, Einstein dismissed ether: "The introduction of a "luminiferous ether" will prove to be superfluous in as much as the view here to be developed will not require an "absolute stationary space " provided with special properties (. . .) (Einstein, 1905). But Einstein soon realized that while he had dismissed the ether of Maxwell and Lorentz, he needed some kind of new ether. In his 1920 lecture *Ether and the theory of relativity*, he stated: "However, closer examination shows that the special theory of relativity does not require an unconditional rejection of the ether". (Einstein, 1976). It requires only that ether should have no mechanical properties at all. As to general relativity, it considers "empty space" as neither homogeneous nor isotropic; hence space becomes a physical medium, which nothing forbids to call ether – of course in a new sense. "The general theory of relativity equips space with physical properties; thus, in this sense the ether exists".

Of course there is another interpretation of general relativity, which goes back to Poincaré's ideas about geometry: it says that the geometry of space-time is a convention that can be freely chosen. (See, for instance, Roxburgh and Tavakol, 1978). Strictly speaking, an experiment never determines the geometry of space-time, only the behaviour of matter in space and time. For instance, Gauss' famous measurement of the sum of the angles of a triangle defined by three mountain peaks can be interpreted as a test of the Euclidean or non-Euclidean character of the geometry of physical space. But it can also be said that it is an experiment about the propagation of light rays.

I will not discuss here this conventionalist interpretation. The conclusion of conventionalists may be excessive. But at any rate, we should remember that we can never study the properties of space (or time, or space-time) "directly"; we have access to them only through the spatio-temporal properties of matter. As far as I know, this idea goes back to Leibniz (Alexander, 1965).

14. An essential contribution of quantum physics to the problem of space is related to the theory of quantum fields. We shall discuss here only quantum electrodynamics, which will be enough for our purposes. The ground state of the free electromagnetic field is an eigenstate of the number of photons, with eigenvalue zero. It is homogeneous and isotropic, or equivalently it is an eigenstate of the generators of the Poincaré group (energy, momentum and so on) with eigenvalue zero. But it is not an eigenstate of the smeared field operators. Therefore, although the vacuum expectation values of these operators are zero, their standard deviations are not zero: the fields fluctuate around their zero expectation value. If we consider now, instead of the free electromagnetic field, the interacting electromagnetic and electronic fields, the same conclusions hold true. Vacuum is characterized by the absence of photons and electrons; but the electromagnetic and electronic fields fluctuate around their zero expectation values.

To describe the coupled fluctuations of these fields, we may speak of virtual particles. We shall say, then, that vacuum is characterized by the absence of *real* photons and electrons, but that in some sense it contains *virtual* photons and electrons, or more exactly: virtual photons and virtual electron-positron pairs, mutually coupled by the electromagnetic interaction.

This description has been very fruitful. In the heroic period of quantum electrodynamics, it allowed the prediction of the effect of vacuum polarization on the spectrum of the hydrogen atom (Uehling, 1935). Even after the discovery of powerful formalisms (Feynman and others), it supplied pictures of great help for physical intuition (Welton, 1948). On the other hand, the theoretical status of this description is unbelievably confused and uncertain. For some physicists (for instance Thirring, Wheeler and others), one may speak about the vacuum and its virtual particles as about any quantum system, without any special caution. For Rosenfeld, on the other hand, virtual pairs and all that are nothing but "easily remembered semi-qualitative features of the formalism" (Rosenfeld, 1959). For most physicists, there is of course no question at all: the formalism works, it gives results in agreement with experiments, what else could we ask? A striking expression of this popular attitude has been given by Wightman in his paper about the Dirac equation (Wightman, 1972, p. 100): in the charge symmetric theory of electrons and positrons, he writes, "the infinite sea of negative energy electrons has vanished from the theory except as a poetic description of the prescription for forming the electromagnetic current". Who cares about poetic descriptions?

15. One of the most striking examples of the uncertain status of vacuum is related to spontaneous electromagnetic radiation. There has been for some sixty years a discussion: can the respective contributions of vacuum fluctuations and radiation reaction to spontaneous radiation be unambiguously separated, or is such a separation merely a matter of convention? The explanation by vacuum fluctuations has a strong intuitive appeal: it allowed, for instance, Welton (1948) to give a simple semi-classical description of the Lamb shift. On the other hand, Fermi (1932) and Ginzburg (1983) remind us that if we consider an atom and the electromagnetic field as two coupled classical oscillators, there is a flow of energy from the first oscillator to the second one both when the latter is already in motion (induced radiation) and when it is not (spontaneous radiation). Hence, according to Ginzburg, spontaneous radiation should not be considered as a quantum effect, because it is not absent in classical theory.

Another more balanced, but not very satisfactory point of view is that the respective contributions of the vacuum fluctuations and the electromagnetic self-interaction to spontaneous radiation cannot be unambiguously separated, because of the arbitrariness of the ordering of non-commuting operators in the interaction terms. This opinion was again expressed recently, for instance, by Milonni et al. (1988). However, already in 1982 Dalibard et al. (Dalibard, Dupont-Roc, Cohen-Tannoudji, 1982) had given a very convincing solution to the whole problem. The right-hand side of the Heisenberg equation of motion for an atomic observable, they argue, is the sum of two terms :

$$\frac{\mathrm{d}G'}{\mathrm{d}t} N(t)E(t) = N(t)[E_{\mathrm{f}} + E_{\mathrm{s}}],$$

where $N(t)$ is an atomic operator, $E(t)$ is a field operator, E_{f} is the vacuum field and E_{s} the field generated by the atom. Now $N(t)$ commutes with the total field operator $E(t)$, but not separately with E_{f} and E_{s}. Therefore, while $\frac{\mathrm{d}G}{\mathrm{d}t}$ does not depend on the order of the operators N and E, the two terms into which it can be separated do depend on this order, hence the ambiguity. Now the main point of Dalibard et al. is that if we want both terms of $\frac{\mathrm{d}G}{\mathrm{d}t}$ to have a physical meaning, they must be Hermitian, as well as the field and atomic operators appearing in their expressions. As a result of these hermiticity requirements, the order of the field operators is uniquely prescribed (it must be completely symmetrical). It turns out, then, that for the ground state of the atom the contributions of radiation reaction (a classical effect) and of vacuum fluctuations (a quantum effect) exactly balance; of course that is not the case for the other atomic states. A purely classical treatment could not explain the stability of the ground state.

One may wonder whether the persistent uncertainty about the physical explanation of spontaneous radiation is not related to the strange status of vacuum. While vacuum fluctuations are taken seriously by Dalibard et al., they are often implicitly considered as mere images, to be used only for the sake of illustration. As is well known, the formalism allows one to compute, for instance, the Lamb shift without any reference to electromagnetic vacuum or to the polarization of electronic vacuum. When the computation is over, we may give as an exercise for students to check that Welton's interpretation, or Uehling's computation, give roughly the right values. It is understood that students may like it, but that mature physicists do not need such naive pictures. No wonder that for many physicists, the above mentioned ambiguity in the explanation of spontaneous radiation was not a real problem. (The situation with the Casimir effect is exactly analogous.)

When one thinks of all the physical effects which have a relation with vacuum, however, it is difficult to admit that there is nothing more than "properties of the formalism", that the physical descriptions of vacuum (à la

Uehling, Welton, Casimir and others) gave correct results by mere chance. And we might add yet other examples. Think, for instance, of the quantum electrodynamics of strong fields (Reinhart, Greiner, 1977). Or of the Unruh effect (Unruh, 1976; De Witt, 1979). The latter is not yet, properly speaking, an "effect", because it has not been seen in an experiment. But it is convincingly proved that if a detector has a uniformly accelerated motion through a vacuum, it does not see an absence of particles (of photons, say), but a thermal bath of photons, with a temperature proportional to the acceleration. (This is a simplified description: to be more precise one should say that the energy of the bath is negative, or else that the photons are not absorbed, but emitted by the detector).

Vacuum and the Interpretation of Quantum Mechanics

16. All these problems with vacuum are difficult, and of course I will not solve them today. I would only like to propose a general idea which might perhaps help to solve them some day. I think that the uncertain status of vacuum is directly related to the indifference of most physicists to the meaning of quantum theory. When Niels Bohr dealt with the problem of quantum theory, he considered it as a true scientific theory – i.e., not only as a formalism equipped with rules of correspondence with the experimental data, but as a full-fledged theory, a theory which can be understood. But now the problem of the meaning of quantum theory is considered by physicists as a field of specialization among others (Bell inequalities, their experimental tests and all that). It is considered as unessential, or even as unsolvable. The trouble with vacuum is but part of the trouble with quantum theory, which is itself but part of the trouble with science in general: we have betrayed the ideal of intelligibility of the founding fathers. We have now a pragmatic science which gives more and more practical results, but which is less and less understood.

As far as quantum mechanics is concerned, the remedy to this illness can be found, I think, in a return to Bohr. Let me first quote an essential corollary of what he called the quantum postulate. "The quantum postulate implies that any observation of atomic phenomena will involve an interaction with the agency of observation not to be neglected. Accordingly, an independent reality in the ordinary physical sense can neither be ascribed to the phenomena not to the agencies of observation". (Bohr, 1934).

That is a statement of Bohr's at the Como conference (1927). The first sentence became very popular under the name of "uncontrollable interaction" between the quantum object and the apparatus (a terminology due to Bohr himself). Perhaps even too popular. At any rate, Bohr criticized it in his 1949 report *Discussion with Einstein on epistemological problems in atomic physics* (Bohr, 1958). In 1938, he writes, he commented again Einstein's view about the incompleteness of the quantum mechanical description. "In this connection I warned especially against phrases, often found in the physical literature, such as "disturbing of phenomena by observation" or "creating physical attributes to atomic objects by measurements". (...) As a more appropriate way of expression I advocated the application of the word *phenomenon* exclusively to the observations obtained under specified circumstances, including an account of the whole experimental arrangement".

Thanks to the recent development of several new experimental techniques, Bohr's lucidity about that very point may be vindicated soon. Indeed, Scully et al. (Scully, Englert, Walther, 1991) have proposed an experiment of atomic interferometry which makes it possible to determine which path the atom follows without disturbing it appreciably. Exit the uncontrollable interaction! But the very fact that the experimental arrangement is appropriate to the acquisition of the "welcher Weg" (which way) information makes the interference fringes vanish. Thus, Bohr's deep idea that what matters at last is not the *perturbation* of the phenomenon but its very *definition* is about to pass a most direct experimental test. As far as I know presently (June 6, 1991), an experiment of the type proposed by Scully et al. has not yet been performed; but the first experiments in atomic interferometry have already been reported (Carnal, Mlynek, 1991; Keith et al., 1991).

Bohr's interpretation of quantum mechanics is directly relevant to the general problem of physical space and time. The recent experimental confirmations of this interpretation should therefore be considered, I believe, as giving even more authority to his point of view about the spatial and temporal character of quantum phenomena. Namely, according to him the space-time picture of physical phenomena is no longer an *a priori* necessity in the quantum domain; it becomes a particular mode of description, which can be achieved only in an appropriate experimental frame.

It is very difficult to adopt the attitude of mind defined by Bohr and to keep it consistently. It is difficult, for instance, to stick to his definition of *phenomenon* (which, by the way, seems to me rather phenomenological in the sense of Husserl). This requires, as Bohr said, a "radical revision" of our mental habits, the necessity of which is not yet perceived or accepted by many physicists. But what can we do else, if really "nobody understands quantum mechanics"? If we really want to understand, we should also want the means of understanding.

17. Bohr's corollary of the quantum postulate also sheds light on the problem of vacuum, as can be seen for instance on the example the Unruh effect. Let us indeed apply the corollary to the vacuum: it says that we must not speak about the vacuum *per se*, but only about vacuum as a part of a phenomenon, observed under specified circumstances.

The temptation to ignore the quantum postulate is very strong here, because of the extreme simplicity of the system of interest: it is *nothing*, or more exactly it is defined by the absence of particles. Or, if we prefer another language, by the absence of real particles. But all these definitions, if we take them literally and forget the quantum postulate, make the Unruh effect something quite odd. If there is nothing, how can an accelerated detector see something? If there are no particles, how can it see particles? (Or, more exactly, a deficit of particles; but this is no less paradoxical). If, finally, we characterize virtual particles as short-lived ones, arising from a provisional violation of the law of conservation of energy (thanks to the fourth Heisenberg relation), then how can these elusive particles appear to an accelerated detector as real ones, or else as a deficit of real ones?

Another type of question has to do with the notion of state. The vacuum is a pure state. But the state seen by the accelerated detector is a mixed one. If the state is an objective characteristic of the system *per se*, how can a pure state be seen by a certain detector as a mixed one?

All these paradoxes are cured by the quantum postulate. Vacuum is a particular quantum system. As such, it cannot be described *per se*, but as an inseparable part of a phenomenon. Then vacuum with a fixed detector is a certain phenomenon, while vacuum with an accelerated detector is another phenomenon. The state describes the phenomenon, not the system *per se*. This is not because we are unable to describe the system *per se*, but because the very notion of system *per se*, borrowed from classical physics, does not make sense any more in quantum physics.

My conjecture is that all difficulties and paradoxes pertinent to the notion of vacuum might be solved in a similar way, by an application of the quantum postulate. But contrary to Rosenfeld, I do not think that notions like virtual pairs have no real value. In a paper about hadronic vacuum (Grandpeix, Lurçat, 1985; see also Lurçat (F.), 1968 and 1977), I described this vacuum with an infinite set of statistical operators, in a way inspired by the works by Dirac (1934) and Heisenberg (1934), but using more versatile mathematical techniques (harmonic analysis of distributions on the Poincaré group, cf. Bonnet, 1984). It turned out that there is a relation between, say, the distribution which describes the virtual pairs of vacuum and the inertial motion of a particle, and similarly between the description of systems of n virtual particles and the processes "p particles give q particles" $(p+q = n)$. (The case of virtual pairs had been sketched by Frenkel (1949) and Thirring (1958)). Thus what is valid in Rosenfeld's objections to Thirring is obviated, because while a system of virtual particles is not a phenomenon, a process is one; hence one may speak about virtual particles, and even describe them precisely, while remaining faithful to the postulate.

Quantum Physics and Phenomenology

18. One cannot say that the name of Husserl has been completely ignored by physicists. (See Weyl, 1952 and 1963; London, Bauer, 1939; Ehlers, 1971; von Weizsäcker, 1973). But there is, I believe, a deep relationship between Husserl's phenomenology and Bohr's quantum postulate, a relationship which as far as I know escaped up to now the attention of both physicists and philosophers.

However, Roman Ingarden, in his contribution to a conference held in 1963, described the constitution of the physical thing according to Husserl and called attention to the similarity between Husserl's ideas and quantum mechanics (Ingarden, 1964). I do not know whether his conjecture of an explicit influence of Husserl's ideas on the development of quantum mechanics (especially in Göttingen) is founded. He pays no special attention to Bohr, but considers quantum mechanics in general. But as far as I know, nobody came as near as him to what I would like to submit now: that Husserlian phenomenology can help to understand Bohr's thought. It does not mean that Bohr was a phenomenologist (he did not care much about philosophical systems anyway). Neither does it mean that I ignore the weak points of Husserl's attempt. (Perhaps phenomenology matches the needs of natural sciences better than those of human sciences.) Let us now come to the point.

Remember the corollary of the quantum postulate (§ 17). The first of the two quoted sentences deals with observation: any observation of atomic phenomena involves an interaction, etc. The second one is a deduction from the first one: "Accordingly, an independent reality cannot be ascribed, etc". This is already phenomenology! According to classical thought, as well as to the common sense of physicists, reality is reality and observation is observation; from a particular feature of observation, we cannot deduce anything about reality. More precisely, if it turns out that observation refuses to answer some of our questions, we shall never conclude from this that our questions were not relevant; instead, we shall try to find other, more powerful observational techniques. Bohr's deduction means that he has already carried out the radical revision demanded by Husserl (whether or not he has ever heard of Husserl's ideas, or read a single line of his works). To draw conclusions about physical reality from an analysis of physical observation means that one uses a "method of analysis of the essence in the sphere of immediate evidence" – and that is precisely Husserl's characterization of phenomenology in its widest sense (Husserl, 1950).

19. Husserl denies the opinion "that the appearing physical thing is an illusion or a faulty picture of the "true" physical thing as determined by physics" (Husserl, 1982, § 52, p. 97 of the original text). He denies the "widely accepted" realism, namely the statement that "The actually perceived (. . .) should (. . .) be regarded as an appearance of (. . .) something else, intrinsically foreign to it and separated from it". (loc. cit.)

For him, "*in the method of physics the perceived physical thing itself* is always and necessarily *precisely the thing which the physicist explores and scientifically determines.* (Ibid., p. 99, modified translation). What he wants above all to refute is the "picture theory" or the "sign theory", namely, that the appearances are pictures or signs of the true real things. A picture or a sign refers indeed to something lying outside it: "in themselves (they) do not "make known" the designated (or depictured) affair itself". On the other hand, "the physical thing as determined by physics (. . .) is nothing foreign to what appears sensuously "in person"; rather it is something which makes itself known (...) in it, and (...) only in it".

This statement – that the physical thing makes itself known in what appears to our senses, and only in it – could be considered, I believe, as a summary of Bohr's interpretation of quantum mechanics; to put it another way, it is equivalent to Bohr's above-quoted definition of a quantum phenomenon. Clearly, Husserl's phenomenology, including the theory of the constitution of the physical thing, can help us to understand Bohr's conception of quantum mechanics. In my book *Niels Bohr* (Lurçat , 1990) I have described, in a popular style, other connections between Bohr and phenomenology.

This does not mean, of course, that Husserl is the only relevant philosopher for an effort towards understanding quantum mechanics. Kant, for instance, should also be considered (see Petersen, 1968, and above all Chevalley, 1991). The usual way of thought of physicists has been a failure in this field. The two great theories of physics that have renewed our understanding of space and time in this century, namely relativity and quantum mechanics, cannot do without philosophy if they are to become genuine sciences, i.e. to be, at last, really understood.

References

Alexander, H.G. (ed.), 1965, *The Leibniz-Clarke correspondence*, Manchester University Press.

Bailey, J. et al., 1977, Nature *218*, 301.

Bergson (H.), Einstein, A. et al, 1922, Bulletin de la Société Française de Philosophie *17*, 91.

Bohr, N., 1934, *Atomic Theory and the Description of Nature*, Cambridge University Press. Cf. *Collected Works*, vol. 6, North-Holland, Amsterdam, 1985, p. 113.

Bohr, N., 1958, *Atomic Physics and Human Knowledge*, Wiley, New York.

Bonnet, P., 1984, Journ. Functional Analysis *55*, 220.

Buytendijk, F.J.J., 1957, *Attitudes et mouvements*, Desclée de Brouwer.

Carnal, O., Mlynek, J., 1991, Phys. Rev. Lett. *66*, 2689.

Chevalley, C., 1991, Introduction and Glossary of: Bohr, N., *Physique atomique et connaissance humaine*, Gallimard, Paris, coll. Folio Essais.

Dalibard, J., Dupont-Roc, J., Cohen-Tannoudji, C., 1982, J. Physique *43*, 1617.

Daniels, N., 1972, Journ. Philos. of Science, 219. (Quoted by Toth, 1977).

De Witt, B., 1979, in: Hawking, S.W., Israel, W. (eds.), *General Relativity*, Cambridge University Press.

Dirac, P.A.M., 1934, Proc. Camb. Phil. Soc. *30*, 150.

Ehlers, J., 1971, in: d'Espagnat, B. (ed.), Scuola Internazionale di Fisica "Enrico Fermi" (Varenna, 1970), Academic Press, New York, vol. 40, p. 478.

Einstein, A., 1905, Ann. Phys. *17*, 891. English translation in: Einstein, A. et al., *The Principle of Relativity*, Dover, New York, 1952.

Einstein, A., 1976, *The inadequacy of classical models of aether*, translation of a lecture given in Leyden, May 5, 1920, in: Capek, M. (ed.), *The Concepts of Space and Time*, Reidel, Dordrecht, p. 329.

Fermi, E., 1932, Rev. Mod. Phys. *4*, 87.

Fock, V.A., 1947, *The systems of Ptolemy and Copernicus in the light of general relativity theory* (in Russian), in: *Nicolaï Copernic*, Izd. Akad. Nauk SSSR, Moscow-Leningrad.

Frenkel, Ya.I., 1949, Doklady Akad. Nauk SSSR *64*, 507.

Ginzburg, V.L., 1974, *Heliocentric system and general relativity theory, (from Copernicus to Einstein)* (in Russian) in: Einsteinovskii Sbornik, 1973, Izd. "Nauka", Moscow, p. 19.

Ginzburg, V.L., 1983, Usp. Fiz. Nauk *140*, 687; Soviet Phys. Usp. *26*, 713.

Grandpeix, J.Y., Lurçat, F., 1985, *Theory of Strong Interactions*, II, *Hadronic Vacuum and Inertial Motion*, Preprint LPTHE Orsay 85/6, unpublished.

Gunter, P.A.Y., 1969, *Bergson and the Evolution of Physics* University of Tennessee Press, Knoxville.

Hafele, J.C., Keating, R.E., 1972, Science *177*, 166 and 168.

Heisenberg, W., 1934, Z. Phys. *90*, 209; 92, 692.

Husserl, E., 1940, in: Farber, M. (ed.), *Philosophical Essays in memory of Edmund Husserl*, Harvard University Press, Cambridge, Mass.

Husserl, E., 1947, *Phenomenology*, in: Encyclopaedia Britannica, vol. 17, p. 699. (First publication 1929).

Husserl, E., 1950, *Die Idee der Phänomenologie*, Husserliana, Vol. 2, Martinus Nijhoff, Haag, p. 14. French translation: *L'idée de la phénoménologie*, PUF, 1970, p. 117.

Husserl, E., 1954, *Die Krisis der europäischen Wissenschaften und die transzendentale Phänomenologie*, Husserliana, Vol. 6, Martinus Nijhoff, Haag. French translation: *La crise des sciences européennes et la phénoménologie transcendentale*, Gallimard, Paris, 1976.

Husserl, E., 1973, *Ding und Raum*, Husserliana, Vol. 16. French translation: *Chose et espace*, PUF, Paris, 1989.

Husserl, E., 1974, *Formale und transzendentale Logik*, Husserliana, Vol. 17. French translation: *Logique formelle et logique transcendentale*. PUF, Paris, 1987.

Husserl, E., 1982, *Ideas pertaining to a pure phenomenology and to a phenomenological philosophy*, 1st Book; E.H., *Collected Works*, Vol. 2, Nijhoff. French translation: *Idées directrices pour une Phénoménologie*, Gallimard, Paris, 1950.

Husserl, E., 1984, Philosophie, N°1, p. 3.

Husserl, E., 1989, *La Terre ne se meut pas*, Ed. de Minuit, Paris.

Ingarden, R., 1964, Archives de Philosophie *27*, 356.

Keith, D.W., Ekstrom, C.R., Turchett, Q.A., Pritchard, D.E., 1991, Phys. Rev. Lett. *66*, 2693.

Kockelmans, J.J. (ed.), 1987, *Phenomenological Psychology, The Dutch School*, Martinus Nijhoff, Dordrecht; coll. Phaenomenologica, 103.

London, F., Bauer, E., 1939, *La théorie de l'observation en mécanique quantique*, Hermann, Paris. English translation in: Wheeler, J.A., Zurek, W.H. (eds.), *Quantum Theory and Measurement*, Princeton University Press, 1983.

Lurçat, F., 1968, Phys. Rev. *173*, 1461.

Lurçat, F., 1977, Ann. Physics *106*, 342.

Lurçat, F., 1990, *Niels Bohr*, éd. Criterion, Paris.

Lurçat, L., 1984, *Le jeune enfant devant les apparences télévisuelles*, éd. ESF, Paris.

Milonni, P.W., Cook, R.J., Goggin, M.E., 1988, Phys. Rev. A *38*, 1621.

Petersen, A., 1968, *Quantum Physics and the Philosophical Tradition*, MIT Press, Cambridge, Mass.

Reinhart, J., Greiner, W., 1977, Rep. Prog. Phys. *40*, 219.

Rosenfeld, L., 1959, Nucl. Phys. *10*, 508.

Roxburgh, I.W., Tavakol, R.K., 1978, Found. Phys. *8*, 229.

Scully, M.O., Englert, B.G., Walther, H., 1991, Nature *351*, 111.

Straus, E., 1956, *Vom Sinn der Sinne*, Zweite, vermehrte Auflage, Springer, Berlin. (There exists an English translation).

Straus, E., 1989, *Du sens des sens*, éd. Jérôme Millon, Grenoble.

Theocharis, Th., 1989, Nature *341*, 100.

Thirring, W.E., 1958, *Principles of Quantum Electrodynamics*, Academic Press, New York.

Toth, I., 1977, La Recherche *8*, 143.

Uehling, E.A., 1935, Phys. Rev. *48*, 55.

Unruh, W.G., 1976, Phys. Rev. D *14*, 870.

Vessot, R.F.C. et al., 1980, Phys. Rev. Lett. *45*, 2081.

Wallon, H., Lurçat, L., 1987, *Dessin, espace et schéma corporel chez l'enfant*, éd. ESF, Paris.

Weisskopf, V., 1981, Physics Today *34*, n°11, 69.

Weizsäcker, C.F. von, 1973, in: Mehra, J., (ed.), *The Physicist's Conception of Nature*, Reidel, Dordrecht.

Welton, Th.A., 1948, Phys. Rev. *74*, 1157.

Weyl, H., 1952, *Space, Time, Matter*, Dover, New York, p. 319.

Weyl, H., 1963, *Philosophy of Mathematics and Natural Science*, Atheneum, New York.

Wightman, A.S., 1972, in: Salam, A., Wigner, E.P. (eds.), *Aspects of Quantum Theory*, Cambridge University Press.

On Renormalization in Quantum Field Theory and the Structure of Space-Time

Ion-Olimpiu Stamatescu

FESt, Schmeilweg 5, D-69118 Heidelberg and Inst. Theor. Physik, Univ. Heidelberg, Philosophenweg 16, D-69120 Heidelberg, Germany

1. Introduction

Physics research is a more dynamical process than it may appear from the textbook descriptions of various physical theories or from a posteriori foundational analyses in the frame of the theory of science. Insights in the research process itself may give hints about the building up of meaning related to the various theoretical constructions and hence may help identify the constructive component in the semantics of physical theories. The discussion presented here is not an analysis in the field of the theory of science, nor a contribution to the history of physics and surely not a theoretical physics study – but is an essay, a trial to circumscribe a certain moment in the evolution of the ideas associated with the concepts of space and time in modern physics. For this, besides briefly describing the physical problems involved, we shall also try to follow the views of physicists concerned with these questions.

The concept of scientific revolution, as it has been promoted by Thomas Kuhn (Kuhn 1962) and widely discussed thereafter, is surely an important tool to globally differentiate between, and to oppose to each other settled theoretical structures in the evolution of knowledge. As a physicist, however, one has the feeling that the conceptual developments have a much more intricate dynamics, which is surely not contradicted, but is also not really caught by the above concept, or at least, by its common understanding as accumulation of contradictions, overthrow and replacement. Steven Weinberg, for instance, says in his Nobel prize lecture 1979 (Weinberg 1980):

"At times, our efforts are illuminated by a brilliant experiment, such as the 1973 discovery of neutral current neutrino reactions. But even in the dark times between experimental breakthroughs, there always continues a steady evolution of theoretical ideas, leading almost imperceptibly to changes in previous beliefs."

It is this subtle kind of change which, I think, may be important to understand when one is asking the question of the development of the meaning associated with theoretical structures. In particular we shall be concerned here with the understanding of space and time in the frame of the modern physical theories. In fact, present day physics research seems to depart strongly from a fundamental concept of space and time which has anything to do with our everyday classical experience. Quantum gravity, strings, supersymmetry, noncommutative geometry are only some titles for various directions of thinking which acknowledge, however, one common trend: a tendency for unification of space-time and dynamics, of the geometrical and interaction structure of the theory. The classical space-time of daily experience appears thereby not only as an approximation, but in fact as a derivative concept, completely "evaporating" at a more fundamental level. Of course, the theoretical models under discussion are very speculative and it is very hard yet to put them under decisive trial either experimentally or by consistency constraints. Nevertheless they seem to suggest that a change in the physicists' approach to space and time is taking place.

The first steps in the reconsideration of the classical concepts of space and time follow from relativity theory (already special relativity, then, of course, general relativity theory), whose main impact concerns the geometry and the global properties, and from quantum mechanics, which raises the problem of localizability and of the conditions for objectivization. These developments have been the subject of many studies in the epistemology of science. Here we shall be concerned with a third step which begins with the establishment of a consistent, special relativistic quantum theory, namely quantum field theory (QFT), and concerns the continuum structure of space and time. Actually we want to argue that trying to understand this evolution, which already shows up in some aspects of the renormalization of quantum field theories in flat space and of the so-called Grand Unification program of the 1970s, is just as essential as understanding the first two steps in order to obtain a feeling for the present day tendencies in physics research concerning the development of the concepts of space and time. This reevaluation process is not yet completed, surely not as long as we lack a consistent quantum theory of gravity, i.e., a general relativistic quantum theory. This is one of the major research efforts and, as indicated above, it already shows some clearly discernible conceptual shifts, some of which may prove irreversible.

Since they deal with space-time distributions, field theories pertain directly to the concept of the space-time continuum. The renormalization program in quantum field theories, which has to be seen as a constituent part of them, concerns the definition of the theory at short distances – i.e., its very fine space-time structure. The results of the renormalization program suggest an intimate connection between the short distance structure of space-time and interaction, which is a fundamental concept in quantum field theory. This connection is especially apparent in the frame of the so-called quantum continuum limit approach. This aspect seems significant for the physical intuition staying at the basis of the search for unified theories of the fundamental phenomena. The models put forward by the present research in the further development of this program go very far beyond Quantum Field Theories in flat space. While still very speculative, this line of thought may indicate a new trial for a unified picture of nature, characterized by a far going unification of space-time, matter and interaction.

In the next section we shall recall some general aspects of modern physics, to help situate the problem. In Sect. 3 we shall shortly discuss the concept of quantum field and its role for the problem of space-time. In Sects. 4–6 some aspects of the renormalization program related to its status in the frame of the theory and to its significance for the concepts of space and time will be reviewed. In Sect. 7 we shall reproduce some opinion expressed by physicists in connection with points under discussion here. Some conclusions will be presented in Sect. 8. The appendix contains an illustration of the discussion of Sect. 6.

2. Some Aspects of the Modern Physics Paradigm

Quantum theory, relativity theory and field theory are the major building blicks of the frame for the description of the fundamental physical phenomena provided by modern physics. To help situate this scheme I should like to make some short comments about its development and about its present implementation.

The picture presented by classical physics before the turn of the century consisted of a number of closed theories providing a more or less coherent system of laws, principles and concepts, built upon the notion of an immovable space-time, and ruling in a deterministic way all developments involving bodies and fields (the latter being usually understood as wavy excitations of some material support – in extreme, the ether). It has been the subject of many studies to show how the accumulation of both factual contradictions and internal inconsistencies led to the fall of this picture. Here I should like to remember only two points:

— Firstly, many of the substantial inconsistencies and contradictions were known for some time – e.g., concerning the structure of matter (existence and stability of atoms), the thermodynamic properties of the continuum (the "ultraviolet catastrophe") or the non-Galilean symmetry of the Maxwell theory. However, only when new solutions were proposed, which did not belong to the "dominant paradigm" and which were experimentally successful – and therefore accepted – only then one began to call the full picture into question.

— Secondly, the revolution took place not so much in the specific theories (which in general remain valid – as approximations at least) but in their basis itself, that is, in the fundamental assumptions about the description of nature achieved by the physical theories. And it is not easy to judge upon the fundamental character of the original contradictions, before one has understood the implications of the new answers.

I wanted to recall by these observations the difficulty of gaining theoretical insight from the discussion of critical aspects of a theory before one has a minimal assertion of their solutions. On the other hand, surely enough these solutions do not come unmotivated, but are promoted in a process of gradual change of meaning accompanying new conceptual associations, the consideration of new empirical facts and the generation of new ideas (this does not contradict the "discovery" or "invention" character of the act of finding the solution).

To give only one example, I shall refer to one instant in Kuhn's careful analysis of the emergence of the "quantum discontinuity" (Kuhn 1987). Kuhn describes at first Planck's use of the statistical mechanics of ideal gases to obtain the black body radiation law, extending therefore the use of thermodynamics to radiation fields. Then he notes:

"Despite its generally classical nature, Planck's statistical radiation theory did differ from Boltzmann's gas theory in one central aspect."

And after quoting Planck himself:

"... the entry of a new universal constant h... marks an essential difference from the expression for the entropy of a gas. In the latter, the magnitude of an elementary region [in the phase space]... disappears from the final result since its only effect is on the physically meaningless additive constant... The thermodynamics of radiation will therefore not be brought into an entirely satisfactory conclusion until the full and universal significance of the constant h is understood... "

Kuhn continues:

"Implicit in this passage is a subtle but extremely important change of emphasis. Though the difference between gas theory and radiation theory is the physical role played in the latter by a particular choice of cell size, what requires explanation is not the necessity for a fixed size but the "significance of the constant h"... The quantum of action proved "cumbersome and refractory", [Planck] notes, when confronted by his efforts to assimilate it classically."

Kuhn's analysis provides an example of "the gradual change of theoretical ideas" mentioned in the introduction and which we are after here in connection with the space-time concepts. The fundamental changes which led to the new physics are of course quantization (Planck, Einstein, Bohr), relativity (Einstein) and – once the ether hypothesis was discarded – the notion of fields propagating in vacuum. The ensuing developments led to the loss of classical determinism and of the classical particles and wave pictures, of the classical notion of absolute space and time and of action at a distance.

Modern physics is dominated by the paradigm of local, relativistic, quantum field theory which represents thus the implementation of the theoretical scheme mentioned at the beginning of this section. These theories possess an extraordinary potential for the unification of the fundamental phenomena, since:

a) - they provide a unique object, the quantum field, to account both for those phenomena which in some well defined classical limit have a particle appearance as well as for those which in the same limit have field (wave) appearance.

b) - they substitute action at a distance by local interaction and solve in this way the old "forces/bodies" dualism. All fields contribute here with the same right – although taking various roles. So the interaction between matter fields (e.g., electrons) is transmitted by the gauge fields (e.g., photons – light) and vice versa. Since particle creation and annihilation is a fundamental feature of these theories, the kinematical concept of particle loses its primordial character, while the role of the interaction (dynamics) becomes dominant. We achieve in this way, among other things a consistent treatment of the "transmutation" phenomena – but also a new picture of the vacuum.

c) - they helped recognize a fundamental constructive principle – gauge symmetry – which may help develop the mathematical scheme for a unified view of the various phenomena (electromagnetic, weak, strong, gravitation). In general, the treatment and understanding of symmetries is well promoted by the quantum field theoretical formalism.

d) - they introduced a construction – the renormalization group (especially in the sense of Wilson) – which allows one to follow quantum phenomena to smaller and smaller scales and trace back in this way the emergence of the diversity of phenomena as induced by phase transitions from a unified, fundamental interaction (Grand Unified Theory – GUT). Thereby, and in connection with symmetries, an intimate connection between dynamics and the space-time continuum was suggested, which, as mentioned in the introduction, then led to a number of new, yet rather adventurous ideas about the structure of the latter, but which may lead to far reaching conceptual consequences in future theoretical developments.

Unification makes thus the *grandeur* of the local quantum field theories – and it also may bring their fall. Not even the phenomena for which there are quantum field theoretical models (electroweak and strong interactions) have yet been unified in a GUT. But there may be also difficulties of principle, related to the weakest of all four known forces, the gravitation (which for various reasons governs both the very large distance and the very small distance phenomena). Namely, going to smaller and smaller scales we arrive at the Planck length, the distance beyond which any attempt to improve the quantum mechanical localization requires energies which are so high, that the curvature of space-time they lead to prevents any further observation (the Compton wavelength of the particle falls below its rising Schwartzschild horizon). The Planck length thus represents a limit for the validity of our theories, in the same way h and c imply limits for the validity of classical, nonrelativistic physics. Only a quantum theory of gravitation would allow us to understand phenomena around and beyond the Planck scale. Although we have a number of ideas about specific quantum gravitation effects (Wheeler, Hawking), a closed theory – free of contradictions – does not exist, and it may be that it cannot be constructed as a local quantum field theory. In fact, the new models in the physics of fundamental phenomena appear to be developing in competition with the scheme of local QFT. However, no empirically supported solution has emerged so far.

It is not clear, therefore, whether a unified description of the physical phenomena can be achieved in the frame of the scheme of local, relativistic, quantum field theories, whether this scheme misses an essential component or whether it must be abandoned altogether. It is the more interesting therefore to try to follow some of the shifts of meaning concerning primary theoretical constructs which have been promoted in this frame but are apparently well secured beyond it. This may be the case with the concept of space-time continuum.

While classical physics supported a space-time concept as the fundamental description frame, independent of the physical objects described by the theory and of their interactions, this separation, already shattered at the beginning of the century by the advent of general relativity and of quantum mechanics, seems completely lost in the modern speculations concerning a unified theory of all fundamental phenomena. One moment in this conceptual reorientation, which seems important to us and will concern us here, already occurs in the frame of the well established quantum field theory, where it is found in the renormalization program and directly concerns the formulation of the theory as a continuum field theory in space and time.

3. Quantum Fields and Space-Time

A field is essentially a density function, i.e. it represents a space-time distribution of some, possibly "complex" quantity: such as, for instance, the color distribution in a painting – although we shall mostly think of a more abstract quantity, like the electromagnetic field or (in the quantum case) the electron-positron field. A field is a coherent distribution, since the values taken by this quantity at various places are correlated – how, this will be described by the theory. For a classical field, this correlation is rigid – given some initial and boundary conditions, the theory fixes the exact values of the fields everywhere and at any time, hence, trivially, also the correlation between different places. For a quantum field this correlation obtains on the average – fluctuations of the distribution occur, which can be described by a superimposed probability amplitude and must be summed over. In both cases, however, this correlations must comply with special relativity theory, implying that they should not lead to measurable faster than light effects.

Since a field "attaches" to every point in space a value of a given physical quantity it may simultaneously be seen as being itself a property of the space. Starting from the continuity of the electromagnetic lines of force, Faraday conceives "that when a magnet is in free space, there is ... a medium (magnetically speaking) around it." And he considers "this outer medium as *essential* to the magnet" (Faraday). But this suggests that aside from the parametric space-time coordinates used to write down the fields as distributions, there is the possibility of looking for a more "physical" concept of a space-time continuum by following the continuity properties of the the (quantum) fields and their correlations, which are fundamental entities of the theory and support a direct physical interpretation. Modern quantum field theory provides a tool for this, the "scaling" analysis, which describes the way the dynamics of the theory is intrinsically related to the scale of the phenomena under consideration. In connection with the renormalization program this allows us to define the approach to small distances in a way which does not make use from the beginning of the properties of the parametric, coordinate space-time. As long as the continuity properties of the fields are trivially related to those of the parametric space-time on which they are defined, as is the case in the classical field theory as well as in the quantum theory of free, noninteracting fields, this distinction is irrelevant. However, in the physically interesting case of interacting quantum fields this similarity no longer holds: the properties of the theory at large do not rescale trivially at short distances, the continuity properties of the physical quantities described by the theory, the field correlations, are no longer trivially related to those of the underlying parametric space but depend crucially on the interactions of the fields. We claim that starting from the scaling structure of the theory one can attempt to identify "objective" elements of the space-time structure – while the parametric coordinate space-time is difficult to use for objective interpretations because of its high degree of arbitrariness. In general relativity, for instance, one attempts to use light rays, particles or matter waves as indicators for the structure of space and time – see, e.g., (Ehlers 1988), (Audretsch and Lammerzahl 1991). Now, general relativity is explicitly a theory of space and time, since, at least locally, we can represent the interaction (gravity) as space-time properties. In the case of QFT in flat space, however, there is no indication in the premises of the theory that the structure of space-time should interfere with the interaction structure of the theory. Also, the effect concerns not the geometry but the continuum structure of the former. Our point of view is that using the short distance behavior of the field correlations we can probe the properties of a space-time continuum which may in some sense be understood objectively, even if possibly not in the frame of the present theories, which do not include gravitation. Actually, since the development aims at a *quantum* theory of gravitation, the notions of "objective properties" (which can be assigned to an "object" and be established by measurement) need themselves a more thorough discussion if they are to be applied to the fundamental quantum entities and not only to some classical phenomena emerging via involved processes of loss of quantum correlations (the present day notion for this is "decoherence"). Since this discussion is not easily summarized we shall avoid it and restrict ourselves here to the use of the less engaging attribute "physical" instead of "objective" to speak of of that fundamental space-time concept which can be traced through the observable phenomena. Also we shall not attempt here to discuss this space-time concept itself, and we are far from trying a constructive axiomatic approach as is the case in the papers quoted above in connection with general relativity. In a much more modest discussion we shall only try to see what scaling considerations tell us about the question of "approaching" short distances.

4. Quantum Fields Theory and Renormalization

A quantum field theory is given in terms of a number of fundamental local fields. These latter may be related to "observable" particles, such as the electrons and photons in QED, or to certifiable constituents of microscopic particles, which however, like the quarks of QCD (the constituents of protons, pions, and the other hadrons) are not observable as isolated, free particles – in any case they must be pointlike, reflecting the local (causal) character of these fundamental fields and their interactions. The dynamics is succinctly described by a Lagrangean density, an expression involving local constructs in the fields and their derivatives – e.g., products of fields at the same point. The equations of motion for the fields, which in principle allow us to describe everything – from the scattering of particles and waves to the building of atoms – are obtained with help of an "action principle" applied to the integral of the Lagrangean density over the contemplated motions.

One way to reproduce the quantum character is to represent the fields as operators on a Hilbert space and write the equations of motion for these operators (more precisely: operator valued distributions). The correlations then appear automatically as expectation values of these operators in the given states – much in the way we are used to from the usual formulation of quantum mechanics in the Heisenberg picture. Another way, which will stay in the background of our discussion here, is to allow explicitly for the fluctuations of the fields by weakening the action principle so as not to demand an extremal "path" (a field configuration in space-time which minimizes the "action") but instead to allow contributions from various paths (various configurations) weighted by a factor depending on the action (the integral over the Lagrangean density). This weighting damps field configurations with strong variations from point to point and other "undesirable" properties – as described by the various terms in the Lagrangean – but does not eliminate them altogether, as it is the case for the classical theory which chooses only the one configuration (path) which (loosely speaking) maximizes the weighting factor. Correlations between fields can be expressed directly in this way, as sums over contributions from various paths. This (Feynman path integral) formulation seems to preserve the classical picture of the field – but of course, since we allow for fluctuations and add contributions from various paths with (at least in the usual formulation with Minkowski metric) complex weighting factors, we must realize that these are not at all conventional objects! In the corresponding formulation of quantum mechanics, say for the double-slit experiment, we must add the contributions from the particle going through one or other slit, which, of course, cannot be taken as a real physical picture in the classical sense.

Over the renormalization program in QFT hung for a time the discomforting aura of an "equivocal procedure to effect ad hoc repairwork on a sick theory". In fact the renormalization program is a well defined procedure and it has to be seen as a constitutive part of a quantum field theory: without renormalization the theory is not sick but just undefined. For an introduction to these problems see, e.g., (Müller 1990), (Zimmermann 1991).

The theory, as given by the Lagrangean, implies a number of parameters, for instance masses, coupling constants, etc. (the former are characteristics of the particular fields, the latter of the interactions among them). But these parameters need not have a meaning as they stand ("bare" parameters), since they are defined as if the fields were free, however only the complete theory makes sense. To define the theory we must fix these "bare" parameters by requiring the theory to produce correct values for a number of physical, measurable quantities. The "renormalization program" consists in two steps: a "regularization scheme", which mainly concerns the mathematically meaningful definition of the symbolic structure of the theory (the calculation algorithms and limiting procedures for the – usually ambiguous or divergent – expressions for physical observables) and, a set of "normalization conditions" for a number of physical quantities (as many as there are possible "bare" parameters in the theory). Since the renormalization procedure represents a convention, the values of the bare parameters depend both on the regularization scheme and on the chosen normalization conditions. However, every other physical quantity in the theory can then be calculated and does not depend any longer on the special renormalization program. Hence renormalization is only a procedure allowing us to "gauge" the theory with the help of a number of *physical* parameters to fix the freedom represented by the "bare" parameters, and to calculate every other physical quantity in terms of these physical parameters.

In fact we can use this independence of the renormalization procedure to write down "covariance equations" for the physical quantities, which may give further insight into the structure of these quantities and allow us to make phenomenological predictions (remember the "phenomenological opacity" of modern physical theories mentioned in the introduction to this book). Consider for example a theory with one "bare" parameter, a coupling constant g_0, and consider that using a certain regularization scheme we can find a physical quantity, a correlation function $\hat{\Gamma}_g(p^2, g_0, \epsilon)$ which represent a good measure for the coupling between the fields – a candidate therefore for a physical definition of the coupling. In general, as the arguments list shows, $\hat{\Gamma}_g$ may depend on the 4-momentum squared at which the measurement is to be made, on the bare parameter g_0 and on a parameter ϵ introduced by the chosen regularization scheme and which is to be taken to some limiting value (say, 0) to formally reproduce the original, non-regularized and therefore in fact not well defined theory. (Usually ϵ plays the role of a short- distance cut-off, since it is the small distance – the UV – behavior which is ill-defined for products of distributions. As long as ϵ is not taken to its limiting value the theory is well defined and all calculated quantities are finite.) Assume we can measure $\hat{\Gamma}_g$ at some value $-\mu^2$ for p^2 and find a value g, depending, of course, on the normalization point $-\mu^2$:

$$\hat{\Gamma}_g(-\mu^2, g_0, \epsilon) = g(\mu)$$
$$\mu \to \mu' : \quad g \to g' = \hat{\Gamma}_g(-\mu'^2, g_0, \epsilon) \tag{1}$$

Then we can in principle reverse this relation to obtain the bare parameter g_0 as a function of g and μ. This is, however, not always meaningful, since the limit $\epsilon \to 0$ usually leads to divergent (infinite) expressions. Fortunately this is also not necessary, since in fact we can rebuild this dependency directly into the correlation functions eliminating thereby the dependency on g_0 altogether. We arrive in this way at renormalized correlation functions which are no longer dependent on the regularization scheme $\Gamma(p, \ldots, g, \mu)$. Since they represent physical quantities which also should not depend on the normalization condition, we can write for them "covariance equations":

$$\mu \frac{d}{d\mu} \Gamma = 0 = \left(\mu \frac{\partial}{\partial \mu} + \beta \frac{\partial}{\partial g} \right) \Gamma \tag{2}$$

where

$$\beta(g) \equiv \mu \frac{\partial g}{\partial \mu} \tag{3}$$

These equations just represent the condition that varying μ and g accordingly leaves the physical quantity $\Gamma(p, \ldots, g, \mu)$ itself unchanged. The so-called "beta-function" $\beta(g)$ of equation.(3), which enters the (loosely speaking) "renormalization group equations" (2), is very useful to describe the change of the effective strength of the interaction with the scale of the phenomena as indicated by the energy scale μ (one calls the function $g(\mu)$ defined by the above procedure also the "running" or "effective" coupling constant). Notice that one can define a "beta-function" $\beta(g_0)$ also for the "bare" coupling g_0:

$$\beta(g_0) \equiv -\epsilon \frac{\partial g_0}{\partial \epsilon} \tag{4}$$

where the minus sign takes care of the reverse dimensionality of ϵ as compared to μ (since we have defined ϵ as a distance).

5. The Renormalization Program

While the logic of renormalization is straightforward (it is essentially a covariance requirement: independence of physics on parameterization conventions), the procedure itself is highly nontrivial. The reason is that both in the Hilbert space formulation, where we have to do with operator valued distributions, and in the path integral formalism, where we have to integrate over the field fluctuations, we encounter mathematically ill-defined expressions. We all learn that this is a consequence of the "interference" between special relativity and quantum aspects and it is due to the infinite number of degrees of freedom implied by the concept of a quantum field and to the nonlinearity of the interaction. It results in the appearance of infinities (divergences) when we try to calculate physical quantities. Our point of view is, however, that this fact, instead of being seen only as a nasty occurrence, should in fact be considered as bearing significance, since it leads to nontrivial properties at short distances and puts very stringent requirements on theory building:

"[one] reason to take renormalization seriously is that this procedure must be considered as an important step towards our goal of obtaining sensible field theories. It simply isn't the final step, but we will not be able to proceed without it." ('t Hooft 1984)

There are essentially two approaches for coping with the problems mentioned above:

I. *The perturbative approach.* One starts from the *linear*, explicitly soluble theory of free fields and one performs a perturbation expansion in the nonlinear interaction term. Since the nonlinear building up of the effects of the interaction may lead to ill-defined quantities, a renormalization program is needed to identify and systematize the divergences and to give a well defined procedure to obtain meaningful perturbative series for the physical quantities. This is the so-called "old renormalization program" and it was satisfactory worked out in the 1960s. However, perturbation theory itself is expected to provide at the best an asymptotic expansion and in fact the most interesting asymptotically free theories show in the region of immediate interest (distances larger than the hadronic scale) dominant non-perturbative effects. There are physical problems, like those of bound states and confinement, which in principle cannot be handled by perturbation theory.

II. *The lattice approach.* One starts with a problem with a *finite number of degrees of freedom*, defined, e.g., by replacing the parametric space-time continuum by a lattice and hence the fields by a discrete set of variables. At this level the mathematics is well defined and the full, nonlinear theory can be treated, e.g., by numerical analyses (at least in the so-called euclidean formulation which implies an analytical continuation to imaginary time). Then one considers a limiting procedure in which the number of degrees of freedom goes to ∞ (the lattice spacing goes to zero) to reconstruct a continuum quantum field theory. This "quantum continuum limit" corresponds to the so-called "new renormalization program" and was established in the 1970s. It has allowed insight into the non-perturbative effects, including predictions of the masses of the hadrons. Its difficulty lies in obtaining analytic results for the (nonlinear) lattice models which would allow us to consider the continuum limit explicitly (most of the results are numerical, hence we need extrapolation procedures and a way to control them).

In the following we shall make a number of remarks concerning mainly results from the "old renormalization program" (Bogoliubov, Parasiuk, Hepp, Zimmermann, Callan and Symanzik, Gell-Mann and Low, et al.); for a review see (Les Houches 1975). To the extent that these remarks do not explicitly imply the perturbation theory they are generally valid. In the next section we shall consider especially the "new renormalization program" in the frame of which the continuum problem appears more directly.

Because it had to deal with infinite quantities the renormalization program was mistrusted at the beginning. So, for instance, P.A.M. Dirac writes (Dirac 1963):

"I am inclined to suspect that the renormalization theory is something that will not survive the future, and that the remarkable agreement between its result and experiment should be looked on as a fluke".

One of the reasons for the early lack of satisfaction was that (at least in the perturbative renormalization of QED) the "bare" coupling and masses turn out to be divergent, meaningless quantities. As we have seen above, it is in fact not necessary to work with bare parameters at all; we can work only with renormalized correlations and parameters and write only meaningful equations for them:

"Critics ... often attack the renormalization procedure as a whole as being ad hoc and unsatisfactory. Now such a negative attitude I do not share. One reason is that the mathematics is completely rigorous if we replace "physically observable numbers" by "infinite asymptotic expansion series" in one or several expansion parameters called (renormalized) coupling constants. The coupling constants are finite, and all expansion coefficients are finite. The only difficulty is that these expansions will at best be asymptotic expansions only; there is no reason to expect a finite radius of convergence."('t Hooft 1984)

However, here is also a deeper epistemological aspect: do we need to define bare masses and couplings in order to identify in the theory the physical particles we want to describe?

"If we insist on the introduction of a bare mass of the electron and the shift of the mass due to self-interaction as the sign of something objective, we run indeed into trouble, since they would both have infinite numerical values. I think this is what Dirac is calling "illogical" in the above mentioned article, since he is explicitly willing to accept a finite shift of a finite mass. The characterization "illogical" is only understandable if one adopts the view that the mass is a substantial quality of the electron and that the mathematical expression for the mass shift and hence for the bare mass is the sign of something objective."

H.G. Dosch, from which this quotation is taken (Dosch 1991), suggests to see this not as an obstacle to objectification but as an incentive to renounce a substantial description in favor of a functional one:

"In Cassirer's scheme however we have no such difficulties, because the expression for the mass shift is (in a regularized theory) mathematically well defined and therefore a truly objective sign. In renormalized perturbation theory one introduces a perfectly finite renormalized mass, which can be related to experimental data. The concept of the renormalized mass is well determined in this theory, but not as a substantial property, since it is only defined in a theoretical framework and its numerical value depends on the so- called renormalization point. This might be an energy scale at which comparison with experiment is made. Even the renormalization procedure (scheme) is not unique, the choice is mostly made according to computational convenience. The results which can be compared with experiment must of course be independent of these choices. The functional dependence of the mass and other quantities of the renormalization point and independence of observable quantities of this choice leads to a quite powerful tool in quantum field theory, the so-called renormalization group equation."(Dosch 1991)

The "effective" coupling constants of a certain class of theories (among them QCD, the theory of strong interactions) show a peculiar behavior: they vanish in the limit of very high momenta (very short distances). This has brought the denomination "asymptotically free" for these theories, and has far reaching consequences – see, e.g., (Zimmermann 1991). One of them is that these theories are under better control, since in the problematic region of small distances, perturbation expansion itself shows that it is increasingly reliable: no proof, but a consistency check.

"Now we have at infinite energy scale an infinitely accurate theory and it is natural to assume that the small distance structure of the theory determines precisely what happens at larger distances. Thus, we imagine that asymptotically free theories are more than just perturbation series: the sums of these series should be well defined, and unique. The conclusion ... is: we should consider asymptotic freedom as a fundamental new step in the advancement of quantum field theory. It now seems to make sense to consider theories where g is not small."('t Hooft 1984)

Another consequence is the fact that a vanishing coupling leaves no free parameter to be empirically fixed any more! What appears instead in the theory is a typical "scale", a kind of fundamental distance (or energy) which essentially differentiates "small" distances, the weak coupling regime, from "large" distances, the strong coupling regime. This effect then suggests the possibility of a unified theory of all interaction: a simple, highly symmetrical and parameter free theory at very small distances, the build-up with the distance of the strength of the interaction leading through subsequent steps of spontaneous symmetry breaking (taking place at a sequence of scales) to the complex hierarchy of the observed phenomena - see, e.g., (Langacker 1981). Although this explicit program has not led to success, the search for the parameter free "theory of everything" born in this program still remains the regulative idea of present physics research; see, e.g., (Pati 1991).

In general one can say that the necessity of dealing with divergent quantities in a consistent way raises the renormalization program to the role of a constructive tool in theory building:

"To a remarkable degree, our present detailed theories of elementary particle interactions can be understood deductively, as consequences of symmetry principles and of a principle of renormalizability which is invoked to deal with the infinities." (Weinberg 1980)

6. Renormalization and Scaling

The "new renormalization program" was worked out in the 1970s (Wilson, Kadanoff, et al.); see, e.g., (Wilson 1983). It puts the accent on the redefinition of the effective dynamics implied by the theory in accordance with the scale of the phenomena envisaged. By using effective coupling constants and other "running" parameters one achieves at each scale a global accounting for the effects taking place at all smaller scales and hence a picture of the physics described by the theory at the given scale. In this way one can treat problems which involve many length scales with apparently very different dynamics at each scale. This is typically the case for theories of fundamental interactions, since they have to proceed from an (expectedly simple) interaction structure at very small distances and reproduce the whole richness of phenomena at all larger scales.

"The "renormalization-group" approach is a strategy for dealing with problems involving many length scales. The strategy is to tackle the problem in steps, one step for each length scale." (Wilson 1983)

The most used implementation of this program proceeds from the "lattice approach" which allows one to have a well defined theory preserving the full nonlinear structure of the interaction. As already noted above, this is achieved by considering a theory dealing with a set of variables distributed discretely according to some neighborhood ordering and a limiting procedure by which a continuum quantum field theory is reconstructed. A very important point is that the whole program can be defined in the frame of the discrete theories (statistical mechanics models with a finite number of degrees of freedom). Thereby the various continuum theories emerge as unique continuum limits of whole classes of discrete theories (universality classes). This construction proceeds in a euclidean formulation, that is with imaginary time (there are a number of conditions under which the Minkowski and the euclidean formulation are related to each other). For our discussion it is interesting to see how by proceeding in this way we can obtain information about the short distance structure of the (limiting) field theory and how the dynamics directly determines this structure. For a textbook see, e.g., (Itzykson and Drouffe 1989).

Let us start with a statistical mechanics model, given as a collection of variables $\{\Psi_n\}$, $n = 1, 2, \ldots$ for which we define "neighborhood relations" and a probability measure (a Gibbs weight factor) for the possible configurations. A configuration is a possible set of values for the Ψ_ns – very much like the distribution of color on the spots of a TV screen. Usually the neighboring relations are defined by imagining a network of points n over which the variables Ψ_n are distributed – a lattice. We can then read immediately which variables are "neighbors". The Gibbs weight factor itself is taken to include a factor depending on the difference between the values taken by two nearest neighbors, and is so constructed as to assign smaller probability to the configurations where this difference is large. Further factors in the Gibbs factor promote in the same probabilistic way the interaction (e.g., by concentrating the preferred values of the variables around some chosen ones). They introduce a number of dimensionless parameters, the so-called "bare coupling constants". Hence the model can be symbolized by indicating the variables (including the lattice specification) and the Gibbs weights (including the coupling constants):

$$[\{\Psi_n\}, g_0, \ldots] \tag{5}$$

Notice that in general there will be more than one type of variable. We speak of "local models" if each factor of the Gibbs weight only depends on the values of a few neighboring variables. For an illustration see the appendix.

Any observable physical quantity is obtained as the average of a certain function of the configurations over all possible configurations, each contributing with the weight given by the Gibbs factor. One interesting quantity is the correlation between two variables located at different points, defined as the average over the configurations of the product of the two variables. This quantity (as well as more complicated correlations) can be calculated, at least in principle, and in the cases of interest for us takes the form:

$$\langle \Psi_n \Psi_m \rangle - \langle \Psi_n \rangle \langle \Psi_m \rangle \simeq e^{-R/R_0} \tag{6}$$

where the distance R depends in some well defined way on the number of lattice points sitting between n and m and $< \ldots >$ symbolize the averages over configurations with the Gibbs weights. The results from the model are expressed in the "physical" correlation lengths $R_0, \ldots$ determining these correlation functions:

$$[\{\Psi_n\}, g_0, \ldots | R_0(g_0), \ldots] \tag{7}$$

They are functions of the bare parameters of the model, $g_0, \ldots$ and in the cases of interest for us may show divergences at some special values for the later:

$$R_0(g_0) \to \infty, \text{ for } g_0 \to g_0^* \tag{8}$$

Now a correlation length has a well defined interpretation as the distance over which the variables "know" of each other since the quantity in equation (6) would vanish if the values of Ψ at n and at m were uncorrelated ($R_0 = 0$). A diverging correlation length means non-vanishing correlation over infinitely many points. But this suggests the emergence of a continuum quantum field theory in this limit! To make this more precise think of the network of variables as a grid through which we view the "real world", and call the grid spacing ϵ – to be measured in some units of length, e.g., cm. The variables themselves would then be values of a field $\psi(x)$ picked up at the nodes of the grid, $x_n = n\epsilon$ (this is written only symbolically: in more dimensions x and n are vectors). In the "real world" we deal with particles showing correlation lengths (inverse masses) measured also in cm. For instance, an electron has a correlation length of circa 10^{-10}cm. A physical distance in the real world, say a correlation length r_0 is then represented on the grid as R_0 points, with:

$$R_0 = \frac{r_0}{\epsilon} \tag{9}$$

Notice that this ϵ is *not* a parameter of the original model (which contains only dimensionless numbers); it is an "interpretation tool" used to understand the limit $g_0 \to 0$ as a continuum limit. What in fact we do is a renormalization procedure by which we fix the physical parameter r_0 and tune the bare parameter $g_0 \to g_0^*$ such as to approach the situation where $R_0(g_0) \to \infty$:

$$\epsilon = \frac{r_0}{R_0(g_0)} = \epsilon(g_0) \to 0 \text{ for } g_0 \to g_0^* \tag{10}$$

We symbolize this procedure by writing

$$[\{\Psi_n\}, g_0, \ldots | R_0(g_0), \ldots]\Big|_{\text{(cont.lim. (non-pert. renorm.): } g_0 \to g_0^*, \ \epsilon = \frac{r_0}{R_0} \to 0, \ r_0:\text{fixed})}$$

$$\to \{\psi(x), r_0, \ldots\}_{LT} \tag{11}$$

The model symbolized by $\{\psi(x), r_0, \ldots\}_{LT}$ represents a euclidean, continuum quantum field theory; it is *defined* as the limiting procedure symbolized by the left hand side of equation (11). What is interesting to notice in this construction is:

a) nowhere did we have to use properties of the continuum field theory – this latter *follows* uniquely as a result of the renormalization procedure which defines the quantum continuum limit discussed above, and if we are not happy with it we must try another lattice model (there are, of course, heuristic criteria for this).

b) this quantum continuum limit is not trivial, since we must change the bare dimensionless parameter with the scale of the phenomena we are considering, in accordance with the "β-function" introduced in Sect. 4, i.e., here:

$$\epsilon \to \epsilon' = \lambda\epsilon: \quad g_0 \to g_0' = f(g_0, \lambda) \to g_0^* \text{ for } \lambda \to 0 \qquad (12)$$

with a certain function f (which is related to the "β-function"). What is the meaning of the scaling relation (12)? On the one hand it signifies the loss of the free, dimensionless parameter g_0 in the continuum theory and its replacement by a scale, defined with the help of r_0, which characterizes the physical phenomena described by the theory. But, on the other hand, one can reverse this connection and consider the limiting process $g_0 \to g_0^*$ as the *construction of a quantum field theory* without any reference to a parametric continuum space-time, and this construction depends essentially on the interaction structure of the theory! If we accept the argument that it is not from the properties of the parametric space-time that we can derive conclusions about a physical concept of space-time, but rather from the behavior of physical quantities like the fields and their correlations, we see here how the interaction intervenes in determining the short distance structure of the theory $\{\psi(x), r_0, \ldots\}_{LT}$ which is *defined* by a renormalization procedure as the limit indicated in .(11).

Certainly this discussion is oversimplified. Also, in the normal case we do not have explicit solutions on which we can "read" the critical points and see the emergence of the continuum field theory. Even in this case, however, the renormalization group analysis can be used as a tool to test the approach to continuum – see, e.g., (Wilson 1983), see also (Müller 1990) for an introductory discussion and references. Also, in the more dimensional case (e.g., three spatial, one temporal dimension, leading in the euclidean formulation to four dimensions) the lattice can only provide a discrete symmetry – usually, the hypercubic one (rotations by 90°). The continuum limit must therefore also reconstruct the full rotation symmetry, letting the invariant $r = \sqrt{x^2 + y^2 + z^2 + \tau^2}$ emerge as the distance between points. These aspects can be studied in full only in numerical analyses, in which one uses very powerful computers to "simulate" the statistical ensemble itself ("Monte Carlo analyses"). But this means that in realistic cases no analytical proof of the continuum limits exists and the argumentation scheme must provide for the possibility of incorporating numerical results. This is an aspect with much farther reaching consequences, also epistemological, than may appear at a first view, but we shall not start here a discussion about "the unreasonable effectiveness of computer physics" (Dreitlein 1992). Of course, in the cases where the perturbative approach for the target continuum QFT makes sense (and this in a restricted sense is the case for asymptotically free theories) we have the possibility of a consistency check by comparing the model defined as the limiting procedure (11) and a construction based on perturbation expansion in continuum, written symbolically:

$$\left[\psi_0(x), g_0, \ldots; \epsilon | \tilde{r}_0 \simeq \sum a_k(\epsilon) g_0^{2k}\right]\Big|_{(\text{perturb. renormaliz.: } \epsilon \to 0, \ g_0(\epsilon) \to g(\mu), \ \tilde{r}_0 : \text{fixed})}$$

$$\rightarrow \sim \{\psi(x), \tilde{r}_0, \ldots\}_{PT} \qquad (13)$$

Here the starting point is a free field theory with a perturbative calculation scheme (defined with help of an action integral related in a certain way to the Gibbs weight factor of the lattice theory equation (5)) and with a short distance regularization symbolized by ϵ. Notice that the model defined through the perturbative renormalization in (13) is only a reduced theory (as we indicated by the tilde) since the perturbation expansion, as a matter of principle, cannot deal with certain effects. Therefore the comparison between $\{\psi(x), r_0, \ldots\}_{LT}$ and $\sim \{\psi(x), \tilde{r}_0, \ldots\}_{PT}$ is meaningful only in an approximate, limiting way (see also the remarks in the previous section).

The analyses of the quantum field theories of the standard model of the elementary particles, especially of quantum chromodynamics (QCD), the theory of quarks and hadrons, use to a large extent numerical methods based on the lattice formulation of these theories. Since perturbation expansion cannot deal with the most important questions these theories are set up to answer, namely concerning the structure of matter, this approach has been developed to the level of a powerful tool – in fact, the only one we have – to deal with these so-called "non-perturbative" effects. It has led in the last years among others to an increasingly good determination of the masses of the hadrons as compounds of quarks, which is one of the crucial tests that QCD is the right theory. It is also possible to study in this approach the scaling properties of the theory directly, by explicitly implementing the renormalization procedure indicated in (12): rescaling of the distances and tuning of the coupling constant to leave physical quantities unaffected. As an illustration we show in Fig. 1 very recent results from this program (QCD-TARO 1993), giving the values of $\Delta\beta$ as function of β for rescaling by a factor $\lambda = 1/2$ (see (12)), where:

$$\beta = 6/g_0^2, \quad \Delta\beta = \beta' - \beta = 1/{g_0'}^2 - 1/g_0^2 \qquad (14)$$

The solid line at the top of the figure gives the value of $\Delta\beta$ which can be calculated perturbatively and should represent an increasingly good approximation for small g_0 (large β), since QCD is an asymptotically free theory. The points represent results from Monte Carlo analyses on four-dimensional lattices containing up to one million points. We see a tendency for agreement between the perturbative (continuum) and non-perturbative (lattice) results at small g_0 (large β), which is a consistency test for the continuum limit, since here we probe small distances. We also see that for not-so- small g_0 the non-perturbative effects are dominant and perturbation theory can no longer be used to obtain predictions from the theory which can be experimentally tested. These non-perturbative effects, however, are most characteristic for QCD, since they describe among other things the structure of matter (the masses of observable particles and what we call confinement). They are dominant at distances equal to or larger than the radius, say, of the proton, since with increasing distance the effective coupling increases (in the lattice model this is reproduced to a certain extent by choosing larger g_0). It is very important therefore that we can deal with these non-perturbative effects, and recent calculations for the masses of observable particles give good agreement with experimental values – see, e.g., (Ukawa 1993). We seem to have here therefore an approach which is not only of abstract interest, but in fact appears to be essential for quantitative studies.

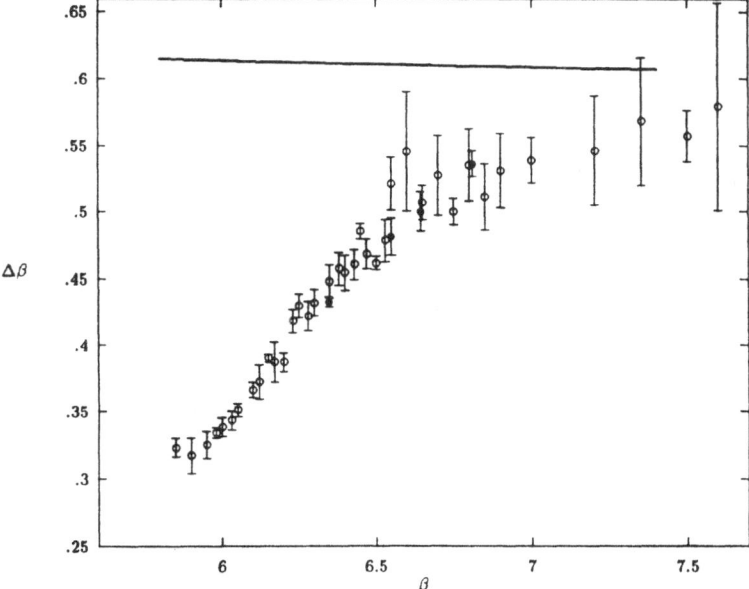

Fig. 1. Testing the continuum limit of lattice QCD. Renormalization Group Analysis: the change of the coupling constant under a reduction of the scale by a factor 2 plotted versus the coupling constant (in fact, the inverse square coupling constants are used $\beta = 6/g_0^2$).

7. Some Views

The continuum concept has a long and rich tradition in mathematics as well as in physics and in natural philosophy. Since we are interested here, however, only in trying to follow the way modern physics approaches this concept in connection with space and time we shall restrict ourselves to quoting some views which we think may help in tracing the course of this line of thought.

Remembering that Leibniz was one of the thinkers who promoted both the calculus and a relativistic view of space as the relations between bodies, it is interesting to quote two of his statements directly pertaining to the spatial continuum. The first one, from *Specimen dynamicum* (1695), is:

"... Woraus offenkundig is, daß der Drang zweifach is, und zwar sowohl elementar, das heißt unendlich klein, welchen ich auch Anregung nenne, als auch durch die Fortsetzung oder Wiederholung der elementaren Dränge gebildet, das heißt der Antrieb selbst. Doch möchte ich deswegen nicht behaupten, daß diese mathematischen Wesenheiten tatsächlich so in der Natur gefunden werden, sondern lediglich, daß sie zur Anstellung genauer Schätzungen durch Abstraktion des Geistes nützlich sind." (Leibniz 1982)

Apparently in contradiction with, he says in a letter to Foucher (Leibniz 1960):

"Je suis tellement pour l' infini actuel, qu'au lieu d'admettre que la nature l'abhorre comme l'on dit vulgairement, je tiens qu'elle l'affecte partout, pour mieux marquer les perfections de son auteur. Ainsi je crois qu'il n'y a aucune partie de la matière qui ne soit, je ne dis pas divisible, mais actuellement divisée, et par conséquent la moindre particelle doit estre considérée comme un monde plein d'une infinité de créatures différentes".

The second statement clearly reflects the strong significance Leibniz attached to the principle of continuity in nature. If we bring the first statement in connection to the "calculus", the apparent opposition between the two statements may then indicate a reservation toward a formal instrument, whose features should not be directly taken over to describe the way nature *is*. If this reading were correct, we should understand it as a warning against hypostatizing the properties shown by our description formalism as substantial properties of the things they should describe.

One century later, Gauss writes more directly in a letter to Bessel, 1829, in connection with geometry – quoted in (Inhetven 1993):

"Nach meiner innersten Überzeugung hat die Raumlehre zu unserem Wissen der selbstverständlichen Wahrheiten eine ganz andere Stellung, als die reine Größenlehre, es geht unserer Kenntnis von jener durchaus diejenige vollständige Überzeugung von ihrer Notwendigkeit (also auch von ihrer absoluten Wahrheit) ab, welche der letzteren eigen is: wir müssen in Demut zugeben, daß, wenn die Zahl bloß unseres Geistes Produkt is, der Raum auch außerhalb unserem Geiste eine Realität hat, der wir a priori ihre Gesetze nicht vollständig vorschreiben können."

But in which direction should we search for this "reality"? Independently of any epistemological question about the role of our intuition of space and time, which can only be classical, and independently of whether we take a realistic stance, following, say, Helmholtz, or a functional point of view, looking for the features of a process of conceptual "condensation", in the spirit, say, of Cassirer, we may ask: is there a conceptual necessity in the frame of our theories indicating a simple continuum structure, extrapolated from that of "our daily experience" and formalized as some differentiable manifold? Here are some more views:

"In what direction should we expect the concept of time to evolve? I believe that the history of elasticity supplies a guide. A great textbook of physics of the 1880s made elasticity a central topic. Knowing the values of the two elastic constants of a material, it showed how one could predict for that material the speed of sound, the bending of a beam and the natural frequencies of a vibrator. But where did the physics of that day get the elastic constants? They were fed in from outside the theory, not deduced from inside. Now, a century later, solid-state physics lets us deduce elasticity from Schrödinger's equation applied to a system of electrons and nuclei. However the very derivation shows us that it has no meaning to speak of "elasticity" in the space between the electron and the nucleus. In brief a concept that was primordial and precise but had to be fed in from outside has been replaced by a concept that is secondary and approximate and derived from inside. We have to expect a development no less dramatic in our view of "time"; expect to see "time", not as primordial and precise, but secondary, approximate and derived. If the last hundred years gave us "elasticity without elasticity", may we not expect the next hundred years to show us the way to "time without time"?" (Wheeler 1984)

In fact there seem to be precise reasons for strong revisions in our concepts of space and time, especially in connection with the problems of quantum gravity:

"Curvature oscillations tend to become uncontrollable at short distance scales. Is there a way to "smoothen out" short scale curvatures? In some sense Nature must become regular there. It is suggestive to speculate that space-time might cease to be continuous but becomes "quantized" into some sort of space-time lattice." ('t Hooft 1984)

In the spirit also of the above interpretation of the two quotations from Leibniz, we should take neither this picture, nor the ones based on "string theories" (which in fact do resort to differentiable manifolds, albeit in a more sophisticated way), etc., literally and as absolute – but we should try to identify in them seeds for the progress of our concepts of space and time. Our stopping at this point (representing more or less the "folklore" of the 1980s) is itself dictated by various considerations, the most important being related to the difficulty of discerning stable trends in frontier research. For an overview of the latter see, e.g., the article of Gernot Münster in the present volume.

8. Conclusions

There are at least three steps in the deterioration of the (classical) concept of space-time as an independent element in a theory:

— the appearance of non-euclidean geometries in general relativity, representing the influence of matter on the geometry of space and time;

— the problem of localization in quantum mechanics: the consistency of a description scheme should be testable by measurements, however, we cannot use material probes at small distances; and

— the problem of short distance divergences in QFT, the dependence of the continuum limit on the interaction – the aspect on which our discussion here concentrated.

These points seem to suggest that space and time must be considered as defined in the frame of a theory together with its matter content and its interaction structure, they cannot be *given* before and independently on the latter. Our "daily" classical space and time themselves emerge in modern physical theories as "derivative" concepts – for instance, as classical trajectories in quantum gravity, "decohered" via environmental interaction with matter (in a distant analogy to traces of microscopic particles in a Wilson chamber). The kind of fundamental space-time concept which would be developed in the future unified theory is still open – we are in a situation not unlike that around 1900, when h was already around but one hadn't grasped its implications yet. Now we think that we shall have new physics at distances around l_{Planck} but we do not know what – this can only come from a consistent theory of gravitation. The problem is, to stress this point again, that the concepts "quantum theory", "symmetry", "space and time", "interactions", etc., cannot be used as independent building blocks, to be simply put together in one or another configuration to build a theory. The lesson from QFT is that the constraints acting in this construction are of a very fundamental nature and can strongly affect the concepts developed separately in the frame of partial schemes.

Appendix

For illustration we write down a two-dimensional euclidean model consisting of a toroidal network of real variables $\Phi_{n,m}$ ($n, m = 1, 2, \ldots, N$; $\Phi_{N+1,m} \equiv \Phi_{1,m}$, $\Phi_{n,N+1} \equiv \Phi_{n,1}$) (the network makes explicit the neighborhood relations: $\Phi_{n+1,m}$, $\Phi_{n,m+1}$ are the nearest neighbors, say, to the right, respectively above $\Phi_{n,m}$). To every physical effect each configuration $\{\Phi_{n,m}\}$, weighted with the Gibbs factor, contributes:

$$e^{-\sum_{n,m=1}^{N}\left\{\frac{1}{2}(\Phi_{n,m}-\Phi_{n+1,m})^2+\frac{1}{2}(\Phi_{n,m}-\Phi_{n,m+1})^2-\frac{1}{2}\mu^2\Phi_{n,m}^2+\frac{1}{4}\mu^2g^2\Phi_{n,m}^4\right\}} \tag{A1}$$

The canonical partition function (the prototype "sum over configurations") is then:

$$Z(\mu, g_0) = \int_{-\infty}^{\infty} d\Phi_{1,1} \cdots \int_{-\infty}^{\infty} d\Phi_{N,N}$$

$$\times e^{-\sum_{n,m=1}^{N}\left\{\frac{1}{2}(\Phi_{n,m}-\Phi_{n+1,m})^2+\frac{1}{2}(\Phi_{n,m}-\Phi_{n,m+1})^2-\frac{1}{2}\mu^2\Phi_{n,m}^2+\frac{1}{4}\mu^2g^2\Phi_{n,m}^4\right\}} \tag{A2}$$

which shows explicitly the role of the Gibbs factor as the "probability" of appearance of each configuration, hence its weight in the sum over their contributions. This Gibbs factor gives small weight to (hence, it makes improbable) configurations with strong variations from point to point (the first two terms in the exponent of equation (A1)), preferring smooth configurations. It also tends to concentrate the variables around the values $\pm\frac{1}{g}$ if μ^2 is large, as it becomes evident by rewriting the last two terms as

$$\frac{1}{4}\mu^2 g^2 \left(\Phi_{n,m}^2 - \frac{1}{g^2} \right)^2 + const$$

which thus represents the interaction structure of this model. In the limit

$$\mu \to \infty, \quad g = \text{fixed} \tag{A3}$$

the "normalized variables" $s_{n,m} \equiv g\Phi_{n,m}$ can only take the values ± 1; they are therefore called "spins", and we obtain in this limit the two-dimensional Ising model which is well studied – see, e.g., (Itzykson and Drouffe 1989), (McCoy and Wu 1973) and and to which we shall restrict the following. The correlation functions between two spins at some distance $R \gg 1$ has been evaluated to be:

$$\langle s_{n,m} s_{n+R,m} \rangle \simeq R^{-1/2} e^{-R/R_0} \tag{A4}$$

showing a "correlation length" which diverges at a certain value of g, the "critical point" $g^* = g_c$:

$$R_0 \simeq \left(\frac{g^2}{g_c^2} - 1 \right)^{-1} \tag{A5}$$

for:

$$g \to g_c = \frac{2}{\ln(\sqrt{2} + 1)} \tag{A6}$$

As discussed in Sect. 6, a diverging correlation length means coherence over infinitely many points, which suggests the emergence of a continuum quantum field theory in this limit! Thinking of our network of variables as a grid through which we view the "real world", we introduce the grid spacing ϵ – to be measured in some units of length, e.g., cm – which is *not* a parameter of the original model but represents a tool used to make explicit the limit $g \to g_c$ as a continuum limit. A specific physical distance in the real world, say a correlation length r_0, is represented on the grid as R_0 points and a renormalization procedure should fix the physical parameter r_0 and tune the bare parameter $g \to g_c$ such as to approach the situation where

$$\epsilon = \frac{r_0}{R_0} \to 0 \tag{A7}$$

Clearly in our model this is possible, with g_c given in (A6), and at this "critical point" we obtain a euclidean continuum field theory. In this case the ensuing continuum theory is explicitly known – it is that of a free *fermionic* quantum field! The lattice model defined in equation (A1) will not be discussed here, but we expect it to show some similarities with the Ising model for large μ.

The scaling relation is:

$$\epsilon \to \epsilon' = \lambda\epsilon: \quad g^2 - g_c^2 \to g'^2 - g_c^2 \simeq \lambda(g^2 - g_c^2) \tag{A8}$$

which shows that the quantum continuum limit is not trivial. If the theory were not to have interaction, or if it were a classical problem, this would not be the case. Consider, say, (A1) with $g = 0$ and $\mu^2 = -M^2$; the model now describes a free *bosonic* field with mass M/ϵ, and the continuum limit means the trivial rescaling $M = m\epsilon \to 0$ for $\epsilon \to 0$, according to the "naive" dimension of the mass parameter. This is very different from the situation described by (A8) (notice that g corresponds to a dimensionless coupling G in the continuum – see below – hence following the "naive" dimension argument it should not change at all!). Consider, on the other hand, the classical problem described by the equation of motion (in the euclidean formulation, with τ representing imaginary time $\tau = it$):

$$\frac{\partial^2}{\partial\tau^2}\Phi(\tau,r) + \frac{\partial^2}{\partial r^2}\Phi(\tau,r) + M^2\Phi(\tau,r) - M^2G^2\Phi(\tau,r)^3 = 0 \tag{A9}$$

The lattice model gives:

$$(\Phi_{n-1,m} + \Phi_{n+1,m} - 2\Phi_{n,m})$$

$$+ (\Phi_{n,m-1} + \Phi_{n,m+1} - 2\Phi_{n,m}) + \mu^2\Phi_{n,m} - \mu^2g^2\Phi_{n,m}^3 = 0 \tag{A10}$$

(this corresponds to retaining *only* the most probable configuration, which maximizes the Gibbs factor (A1)). Clearly the continuum (A9) is obtained from equation (A10) by taking the "naive continuum limit":

$$\epsilon \to 0: \quad \tau = n\epsilon, \ldots, \quad \Phi_{n+1,m} - \Phi_{n,m} \simeq \epsilon\frac{\partial}{\partial\tau}\Phi(\tau,r), \ldots,$$

$$M \equiv \epsilon\mu \to 0, \quad g = G : \text{fixed!} \tag{A11}$$

Again, all quantities scale according to their "naive" dimensions, since there is here no integration over fluctuations involved, which may change the power counting.

References

Audretsch, J. and Lämmerzahl, C.(1991), J. of Math. Physics 32, 8, 2099

Dirac, P. A. M. (1963), The evolution of the physicist's picture of nature, Scientific American, May 1965

Dosch, H. G. (1991): Renormalized Quantum Field Theory and Cassirer's Epistemological System, in Proceedings of the 2nd Philosophy-and-Physics-Workshop, Phil. Nat. 28, 1

Dreitlein, J. (1993), Found. of Physics 23, 6, 923

Ehlers, J. (1988), Einfürung der Raum-Zeit-Struktur mittels Lichtstrahlen und Teilchen, in Philosophie und Physik der Raum-Zeit, J. Audretsch und K. Mainzer Eds., BI-Wissenschaft-Verlag

Faraday, Michael, Article 3277 in "Experimental researches in electricity"

Feynman, Richard (1986): The Character of Physical Law; MIT Press, London

Inhetven, R. (1993), Empirie und a priori, Report Nr. 3/93, ZIF, Bielefeld

Itzykson, C. and Drouffe, J.-M. (1989), Statistical Field Theory, Cambridge University Press

Kuhn, T. S. (1962): The Structure of Scientific Revolutions, University of Chicago Press, Chicago

Kuhn, T. S. (1987): Black-Body Theory and the Quantum Discontinuity, University of Chicago Press, Chicago

Langacker, P. (1981): Grand Unified Theories and Proton Decay, Physics Reports 72, No.4, 185-385

Leibniz, G. W. (1982) Specimen Dynamicum, Felix Meiner Verlag, 13

Leibniz, G.W. (1960): Schreiben an Foucher, Philos. Schriften (Gerhardt), I, 416

Les Houches (1976): Methods in Field Theory, Balian, R., and Zinn-Justin, J. eds., North-Holland

McCoy, B. M. and Wu, Tai Tsun (1973), The Two-Dimensional Ising Model, Harvard University Press

Müller (1990): Lokale Quantenfeldtheorie und Kontinuumlimes, in Proceedings of the 1st Philosophy-and-Physics-Workshop, Phil. Nat. 27, 1

Pati, J. C. (1991): Grand Unification in a Broader Perspective, in Proceedings of the A. D. Sakharov Int. Conf. on Physics, Moscow

QCD-TARO Collaboration (1993), QCD on the massively parallel computer AP1000, in Proceeding of Conference "Physics on highly parallel computers", Jülich

't Hooft, G. (1984): Quantum field theory for elementary particles, in Higher Energy Physics, Physics Reports, 104, Nos. 2-4, 129-142

Ukawa, A. (1993), Nucl. Phys. B (Proc. Suppl.) 30, 3

Weinberg, Steven (1980): Reviews of Modern Physics, Vol. 52, No. 3

Wheeler, J.A. (1984) Harbingers of a coming third era of physics, Higher Energy Physics, Physics Reports, 104, Nos. 2-4,

Wilson, Kenneth G. (1983), Rev. Mod. Phys., Vol. 55, No. 3

Zimmermann, W. (1991), Physikalische Blätter, 47, 605

Quantum Theory – a Window to the World Beyond Physics

Euan J. Squires

Department of Mathematical Sciences, University of Durham,
Durham City, DH1 3LE, England

1. Beyond Quantum Theory

It was the science of classical mechanics that made the whole idea of "reductionism" – explaining things in terms of the properties of their constituents – anathema to all who maintain that there is more to the universe then the mere relentless working of automata, eternally governed by Newton's laws of motion. Those laws, of universal applicability, seemed to have no place for mind and spirit, for consciousness, for hope and fear, for love, for meaning, still less for free will or the workings of the divine in the affairs of the world. Of course, this was never really true – certainly Newton would not have accepted such a consequence of his ideas – essentially because the task of science is to explain human experience. The concept of experience is there at the outset; without it there is nothing for science to be about.

Nevertheless, it is possible to do classical mechanics in a way that ignores experience, and deals simply with the world *as it is*. Suppose, for example, I do an experiment in which I take a ball and drop it. Then I see, i.e., experience, the ball hitting the ground at a certain time. I can forget this experience, however, and simply say that *the ball hits the ground at a certain time.* Then I use classical mechanics, together with the law of gravitation, to calculate the path of the ball, and I find that this calculation agrees with what actually happens. I have a theory which explains a particular fact about the world.

The most important discovery of this century is that such a procedure is (almost certainly) in general not correct. (The reason for the qualification in parenthesis will come later.) We live in a quantum world, and the rules of classical physics are an approximation which is adequate only in very special circumstances. (In passing it is worth noting that very likely there is no choice here. The laws of classical physics really do not seem to allow for a world in which large objects of any sort, still less interesting things like us, can exist.) To understand the difference that quantum theory brings we need to consider a slightly different experiment. Here we imagine the ball (which we shall now refer to as a particle, since quantum effects are most readily evident for very tiny objects) moving towards a "barrier" of some sort. We can think of this as being like a particle rolling along a level surface. The barrier represents a hill. According to classical mechanics the particle will roll over a low hill (it will be transmitted), but be turned back (reflected) by a large hill, exactly as will be observed. It is this simple prediction of classical mechanics that turns out to be wrong. We can discover this if we put mirrors in place (see Fig. 1a) to reflect both a transmitted and reflected particle towards the same detector (D in the figure). It turns out that if we do this then we always find that there is a contribution from *both* mirrors. A peculiarly wave-like phenomena known as *interference* is used to demonstrate this inescapable fact. In other words, the statement that the particle either went through the barrier, or was reflected by the barrier, is wrong. The only thing I can safely say is that, if I *observe* it, e.g., as in Fig. 1b, then I see that it either went through or was reflected. We have discovered that there is a mysterious gap between what *is*, and what I experience.

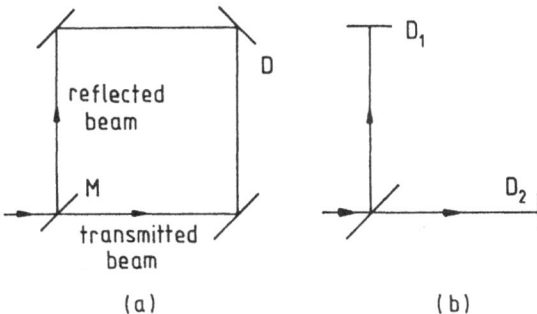

Fig. 1. Here an incident beam is split into two parts by a "half-silvered-mirror" at M. In (a) the reflected and transmitted parts are brought together at the detector (D), where they are seen to interfere. In (b) it is observed that, for a single incident particle, either detector D_1 or detector D_2 detects the particle – never both.

Now I believe that quantum *theory* gives us a partial understanding of what is happening here. (There are some who would say that this is not so, and that quantum theory cannot be applied to a single system. I do not myself think this is a reasonable view. In any case the people who believe it have not come up with an alternative). In quantum theory what we thought was a particle, a tiny object following a well-defined path in space, is really a wave. Then, when the wave reached the barrier it split into two parts, a reflected part and a transmitted part (this is a very familiar wave phenomenon seen when light falls on a pane of glass for example). The fact that both mirrors contribute to the final effect is readily understood. What is not so easy to understand is my observation that the "particle" goes only one way! More generally, quantum theory, in its orthodox form, does not have particles (where these are here defined as small objects that are always at a definite place).

Here we reach what is often called the "measurement problem" of quantum theory. This theory appears to be *correct*; it has dominated the physics of this century and shown itself to be applicable to a wide variety of phenomena, allowing accurate numerical predictions to be made. Nevertheless, it does not have in it any particles (more generally, "events"), i.e., the objects of experience as they are actually observed. This gap between the theory and the experience is bridged by a rule that says that if you make an observation there is a certain probabilty that you will see a particular event, e.g., a particle at a particular place. We emphasise that this cannot be understood by saying that there *is* a particle at that place! In some mysterious sense the observation itself creates the particle.

At this stage it is important to realise that, although our description here has been rather vague, there is a precise mathematical formulation of quantum theory, which allows exact calculations to be made. In many cases, however, the observed results are not produced in these calculations. Instead, what is produced is a so-called "superposition" of many results, and the theory gives the probabilities of what will be seen when an observation is made. When a particular experiment is repeated many times these statistical predictions do indeed turn out to be correct. In the previous example it is the transmitted and reflected objects that are in the superposition. In fact, using the rather formal mathematical language of quantum theory, we calculate that the state after interaction with the barrier and the two detectors (Fig. 1b) has the form:

$$\Psi = \alpha(R, D_1^*, D_2) + \beta(T, D_1, D_2^*),$$

where here α and β are two constants which give the relative amounts of the reflected wave, R, and the transmitted wave, T, respectively, and D_1^* represents the detector D_1 having detected a particle, and similarly for D_2.

An obvious development now is to try to describe, i.e., find a theory for, the act of "observation". The most important thing we know about this is that it cannot be described by quantum theory itself. As long as we assume that the "apparatus" used for observation is quantum-mechanical then we know that we will always keep the superposition, and never get the observed result. In fact we can include a sequence of devices to make the observations, going all the way, for example, to my brain. The appropriate mathematical description of the state will then be:

$$\Psi = \alpha\big(R, D_1^*, D_2, A(R), B(R)\big) + \beta\big(T, D_1, D_2^*, A(T), B(T)\big),$$

where $A(R)$ means the state of the apparatus, A, when the detector D_1 has recorded the particle, $B(R)$ is the corresponding state of my brain, and similarly for the states $A(T), B(T)$ which describe the apparatus and my brain when the transmitted particle has been detected.

According to quantum theory this is the complete description of the state. Nowhere, however, does it contain anything that corresponds to my experience, which I know is one thing or the other; never is it both. Thus we have arrived at an important conclusion: there is something beyond quantum theory. Another way of saying this is that, if quantum theory is the complete description of physics, then "mind-brain identity" theories, or theories involving so-called "psycho-physical parallelism", are wrong.

2. Modified Quantum Theory

Of course everything in the previous section was known, and much discussed, in the very early days of the development of quantum theory. At first there was the possibility of "looking back", i.e., of pretending that classical apparatus really existed. They could be used in the theoretical discussions whenever it was convenient. Such an option is not now available. Ultimately everything has to be understood quantum-mechanically; there are no classical objects. Another frequently used escape is to retreat into some sort of positivism or instrumentalism – the task of science is merely to make predictions, not to seek true explanations. Although ideas of that nature were beneficial in that they created the right sort of climate to allow quantum theory to become accepted and used, they are in my view contrary to what science is about. Our aim is to explain our experience; to tell a convincing story about what is actually happening "out there". Certainly if we give up looking for explanations we will not find them, so some of us, at least, will continue to try.

Incidentally, I would not want this discussion of the difficulties of understanding quantum physics to give any encouragement to what I see as a dangerous, though rather fashionable, "anti-science" movement. This movement tends to unite a justifiable concern for our planet and for human life, with an unjustified blaming of science and scientists for mankind's failure to control the greed and violence that threaten us, and hence to belittle the claims of science as a serious attempt to understand many features of our universe. Neither the cause of understanding, nor of humanity, are served by such anti-science arguments and prejudices.

To return to our problem, we can try to modify the laws of quantum theory, or add to them, so that the gap between what is calculated and what is experienced is filled. Some excellent attempts along these lines exist; they are good physics, they work, and greatly deserve careful study, and they explain the reason for the parenthetical remark in the third paragraph of this article. Nevertheless, the community of physicists does not take them seriously. Partly this is due to ignorance. There are also some good reasons why no extended version of quantum theory is convincing. Quantum theory is a beautiful and simple theory, so it is unfortunate to have to add features which lack the same elegance; the theory is calculationally so successful that care is necessary not to change significantly the predictions, giving the feeling that nature seems to have struggled hard to make orthodox quantum theory give the correct numerical answers.

For many people the most serious objection is that the physics of the new theories is not "local". A local theory is one in which we can discuss experiments in a small region of space and time without worrying about what is happening in other distant parts of the galaxy. An alternative way of saying this is that anything I do here will have an effect at other places which will get less with increasing distance, and will take time to reach other points of space. (Some sort of assumptions along these lines are necessary if we are to do science at all). Now, particularly since the theorem of John Bell in 1964, and the subsequent experiments of Aspect and his group in Paris, we have realised that the quantum world has curious non-local features. These are so important that we shall devote the next section to them.

Of relevance to our discussion here, we note a strange feature of the non-locality. Although it is clearly present in quantum theory, it is, in a curious way, hidden; in particular, it does not seem to allow us to send instantaneous signals. Nevertheless, any modification of quantum theory which "solves" the measurement problem almost inevitably brings this non-locality into the equations. We are very unhappy with equations of this type, particularly as they violate Einstein's relativity principle (see the next section), and hence we are inclined to reject theories which require them.

Work in this area will go on. Clearly it is work in which I, as a theoretical physicist, must remain interested. There is a strong desire among physicists to regard physics as the "theory of everything", and a reluctance to allow for anything to be beyond its scope. Indeed we might try to define it by some such phrase. With so many things not understood, however, such a definition is surely not too sensible at the present stage (it may also run into Gödel-like contradictions), so we may want something more precise. We could for example demand that the laws of physics are just the laws of orthodox quantum theory (this has the advantage that they are then local and not obviously in violation of Lorentz invariance!).

Given the conclusion of the previous section we are then inevitably led to the most exciting aspect of quantum theory, namely that in which it provides a window to a world beyond physics.

3. Non-locality

One of the supreme successes of the application of mathematics to the physical world was Isaac Newton's explanation of planetary motion in terms of his law of gravitation. This brought "the heavens" into the domain of science, and began the process which has reached its present peak in Big Bang cosmology. Newton, however, was not entirely happy with his explanation. The whole scheme was based on the idea that the sun could influence a body with which it was not in contact, and indeed with which it had no obvious mechanical linkage. Such *action at a distance* was, to him, "an absurdity".

It was over two centuries later that Einstein, in his new theory of gravitation (general relativity), provided a partial resolution of this problem. In this theory the effect of gravity propagates as a wave travelling at the speed of light. Hence, although the wave travels in empty space, and Newton would certainly have been puzzled about what is actually doing the "waving", the locality is restored; an event at one point of space can influence what happens at another only because something travels at a finite speed between the two points.

As with his illustrious predecessor, locality was a crucial feature of Einstein's view of reality. Even before he had restored it to gravity, he had put the idea on a much sounder basis in his earlier theory of special relativity, an important feature of which is that no significance can be attached to events at different spatial points being "simultaneous". Thus instantaneous action at a distance becomes incompatible with special relativity. In 1935, along with Podolsky and Rosen, Einstein used the requirement of locality to demonstrate what he regarded as the incompleteness of quantum theory. The basic (EPR) idea is that two particles, in a so-called correlated quantum state, go to opposite ends of some apparatus, where particular properties can be measured. According to orthodox quantum theory what happens at one end can be fixed instantaneously by a particular measurement at the other. Locality forbids this, hence in fact the information must already have been present at the first end. Thus, it was argued, since quantum theory did not have this information, it must be incomplete.

The EPR paper aroused a lot of controversy and initiated a series of letters between Einstein and Bohr. Amazingly, however, a key point was missed, and it was not until 1964 that John Bell showed how the EPR situation could actually be used to demonstrate that *no way of completing quantum theory by adding extra information (hidden variables) could be both local and agree with the orthodox predictions of the theory*. In some sense this showed that quantum theory is intrinsically non-local. Naturally this led to a desire to do some EPR-type experiments. Would quantum theory or locality turn out to be correct? A series of very careful experimental tests, carried out in Paris in the early 1980s, found complete agreement with the predictions of quantum theory, even where these violated the predictions of locality. Most physicists have accepted these experiments as conclusive. In my opinion, non-locality is such a scandal to physics that this is perhaps a little premature, and we should still look for loopholes in the interpretation of the experiments. I have myself suggested some possibilities. Nevertheless, the history of 20th century physics has been that of the ever widening and continuing success of quantum theory. If this has any lessons, it would suggest that quantum theory will remain correct, and that some sort of mysterious action at a distance will have to be accepted.

Since the original Bell theorem, there have been other demonstrations of the peculiar non-locality of quantum predictions. (We all think our own is the simplest.) For completeness, here is another. Imagine you have a set of cards, numbered 1,2,3,..., etc. On each card there is a red and a blue patch, and you are told that you can remove either the red or the blue, but not both. You have a free choice which you decide to remove. Underneath each patch you discover that there is a letter which is always Y (yes) or N (no). You record the letter, along with the number of the card.

You are then told that another person was given a similar set of cards, also numbered 1,2,3,..., so that the cards are naturally paired. This person went through an identical procedure, removing a freely chosen colour and recording a Y or an N.

Now you are told that the cards were made so that when you compare your records the following combinations, and only these, can occur:

Both blue opened: Y,Y or Y,N or N,N.

Both red opened: Y,N or N,N.

One blue and one red opened: Y,Y or N,N or N under the blue and Y under the red.

At this stage you conclude that somebody is trying to fool you. Such cards surely cannot be constructed. This appears to be easy to demonstrate. We merely have to consider the cards that give Y,Y when both blues are opened. Clearly I could have chosen instead to open the red, in which case it is clear from the above that I must have obtained a Y (a blue Y and a red N is claimed not to exist). Equally, of course, the other person could have opened the red when I opened the blue and, by the same argument, a Y would have appeared. It therefore follows that if we both chose to open the red, then these cards must both have revealed Y, in contradiction to what is claimed above as possible for two reds.

If quantum theory is correct then this argument is wrong, and the cards can indeed be constructed to behave as specified. (I am speaking theoretically here. There would be technical difficulties in doing things exactly as I have stated – certainly very thick cards would be required!). Where the argument goes wrong is in its assumption that the letters are already present before the labels are removed. This is a bit like assuming that the particle is already present before it is observed. In fact, the act of removing the label (this is the observation) creates the letter. What is more, and this is where the non-locality arises, the precise letter that it creates depends on what has happened to the other card, i.e., which label has been removed. This remains true regardless of the distance between the two cards when they are examined! I hope it is clear why I said before that this is a scandal for physics. If the world really is like this, then it is way beyond anything we can at present hope to comprehend!

Particularly for those with a knowledge of quantum theory it will be useful to include here the method of constructing cards with the above strange properties. It is due to Lucien Hardy and uses two interferometers with an intersection point as shown in Fig. 2. Each interferometer is tuned so that, if the other is absent, there is complete destructive interference in one direction (a dark output). There is a possibility of placing detectors in these dark output channels (these are the "blue" detectors), and also in the two intersecting paths (these are the "red" detectors). We now imagine that particles enter the two interferometers simultaneously, so that if they both travel along the intersecting paths they annihilate each other. It is a simple exercise in elementary quantum theory to see that we will obtain the above results if we interpret "opening the red patch" as measuring with the red detector, etc., and recording (not recording) a particle as Yes (No).

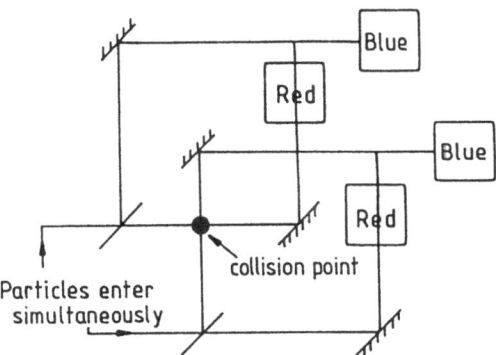

Fig. 2. The Hardy intersecting interferometers experiment, which can be used to give a demonstration of quantum non-locality

4. Beyond Physics

In Sect. 2 we defined "physics" in a particular way, and saw that it did not contain the things of experience, i.e., the particles and events of the observed world. Rather, it simply gave probabilities for such events to occur. Clearly, then, we need something else to actually make the selection from the available options. It is natural to try the idea that this something else is, or is closely related to, consciousness. There are several reasons why we might want to do this. The problem of the nature of consciousness, and how it can be related to the physical world, is notoriously difficult and has been the subject of numerous speculations, with little real progress. An approach that begins from the world of microphysics is at least new, and may produce fresh insight. Also, it is only because of consciousness that we know that our quantum theory is inadequate; it could well be that the laws of quantum theory do indeed correctly describe everything in an unconscious universe, i.e., if it were not for the existence of consciousness there would be nothing beyond those laws.

Finally, there is the criterion of economy. Neither whatever it is that "selects" one event from the range of quantum possibilities, nor consciousness, appear to be contained in what we have defined as physics, so it makes sense to try to make do with one new thing rather than to introduce two.

Here we are following Descartes, who introduced the notion of "dualism" by asserting that reality consisted of two things: matter and soul. Our ingredients are different. The "out there" existing object of quantum physics is not "matter", rather it is the somewhat abstract notion of a wavefunction (sometimes called a state-vector), which evolves in time according to a specific equation (the Schrödinger equation). In addition we have consciousness, and it is through the interplay of the two that what we know as matter appears. This interplay takes place certainly in human brains, and it is reasonable to suppose that it is the enormous complexity and sensitivity of brains that allow it to occur. Thus the model suggests that physical systems are "conscious" to varying degrees. In this sense our model is "panpsychic"; whether something is, or is not, conscious is not an arbitrary decision as to whether it was given a "soul", but rather it is a question of how effective it is as a window between consciousness and the world of physics. We could at least imagine that the degree of effectiveness could be calculated, though this would require that we had some sort of model for the operation of consciousness.

The central function of consciousness as introduced here is selection, and it is possible to argue that, in some sense, this is what consciousness "feels like" (compare recent remarks of Roger Penrose). Certainly we select the focus of our consciousness at any given time. Notice also that it is through this selection that consciousness has an effect on the experienced world. The way this happens has always been a mystery and an acute problem for most theories of the relation between consciousness and the physical world. On the one hand, the latter seems to be closed, i.e., to have laws that determine how it will behave without having any room for outside interference. On the other, it just does not seem to be adequate to regard consciousness as an "epiphenomenon", having no effect whatever on what happens in the world. Somewhere there seems to be an argument that if consciousness did not exist, then the world would be different, e.g., I would not be writing about it. (The same argument of course is wrong if used about "ghosts", or even God, so some care is needed. The point is that it is hard to see how consciousness could be "imagined" by machines that did not possess it – or even what this could mean). With our model there is no problem about how consciousness can be truly efficacious. The complete physical world, i.e., that defined by the wavefunction, is unchanged by consciousness. (This view of quantum physics, in which the wavefunction is the full story, and all the possibilities actually "happen", is nowadays often called the many-worlds interpretation). Nevertheless, the observed world is that which happens to be selected, and here consciousness plays a vital role.

In most cases it is reasonable to believe that the selection is made at random. For reasons we do not understand consciousness has to see, for example,

a particle at a given place, and it selects that place from the available possibilites given by the wavefunction. The standard probability rule of quantum
theory is quite natural, since it involves the obvious choice of the probabilty
weight which is required to give meaning to the term "random".

It is now possible to speculate that, where the superposition of different
possibilities has its origin in the brain of a person, rather than in some outside
quantum system, then the selection need not be made at random, but could
be the subject of conscious choice. Such a speculation could then lead to the
suggestion that it is such a situation that gives rise to the experience of free
will. A possible objection to this type of idea is that it requires the brain to be
operating as a genuine quantum system, i.e., one in which quantum effects are
important, whereas at the level of understood brain activity this cannot be
the case. I am not sure whether this is a serious objection, since at *some* level
the brain, and indeed all "matter", has to be sensitive to quantum effects.

Of course if an idea like this were to be believed, there would be an obvious
further speculation. Why should deliberate choice not be possible even in situations where the quantum superposition originates outside the brain? Could
we not make a deliberate selection of whether or not a particle is transmitted
by a potential barrier, and so obtain results that violate the quantum probability laws? Maybe we could, and the fact that we do not is a consequence
simply of our not having learned how!

5. Beyond My Consciousness

Now we come to what seems to me to be the most significant aspect of the type
of model we are here discussing. We have said that when I make an observation
of a quantum "superposition" I select a particular outcome. This is what I
observe. Suppose, at the same time, or later, you also observe the system.
You too will need to make a selection. Can you also make it at random? If
so then you might make a different selection to me. What would then happen
when we meet? How could we ensure that we would agree about what had
occurred (and normally we expect such agreement)? The first thing to say here
is that, from the point of view of physical systems, there is no problem. It is
a direct consequence of the physics that only the brains which had had the
same experience (belonged to the same branch of the wavefunction in many-
worlds language) could interact. This fact is crucial to the whole idea of the
many-worlds interpretation. However, we have gone beyond physical systems,
and it is surely necessary that our conscious choices agree. In other words I
want the physical you that I meet to be the conscious you. How can this be
guaranteed?

Before giving an answer, it should be noted that this type of problem (not
necessarily in this language) was realised in the earliest days of quantum theory. It was answered by supposing that the first observation somehow changed
the wavefunction (collapsed it), and most textbooks refer to such a collapse,

usually without noting that it violates the laws of quantum theory, in particular the Schrödinger equation. Perhaps this is why Schrödinger himself never believed that the collapse happened. He recognised that there was only one alternative: if the first observation did not somehow "mark" the bit of the wavefunction that it had selected (such a mark would imply a change in the physics), then there could be only *one* selection, which leads naturally to the idea that there is only one, universal, consciousness.

There is, of course, nothing very novel about such a suggestion. What is remarkable is that it has come from consideration of the microworld of physics. It is from our study of the world of elementary particles, from reductionist physics, the attempt to understand everything in terms of its constituents, that we have been led to postulate the idea of a universal consciousness. What properties does it have? First, it determines those things that are not determined by the laws of physics. Hence it plays a significant role in making the world to be as it is. Partially, at least, there may be deliberate choice, "design", operating here. Secondly, it operates in and through human brains, and is the ultimate origin of all "feelings". Thirdly, it transcends the physical restrictions of time and space (recall the discussion of non-locality, and the fact that in our model here the "physics" is local, so that it is consciousness that carries the non-locality). Fourthly, if we move into the domain of quantum cosmology, where the principles of general relativity suggest that the wavefunction of the universe is independent of time, then the first act of universal consciousness, i.e. the first observation, may be regarded as the creation of the universe in that it creates "time" (an idea of which Augustine would have approved).

As Schrödinger noted (and used, somewhat dubiously in my opinion, as evidence) every age and culture has tended to accept the existence of such an entity. Within the Judaic, Christian, Islamic traditions, it is known as God. In our very sad world, where this idea is so often a cause of division, it is worth adding that insofar as there is support for the idea from the world of physics, it is its *oneness* which is crucial. At the heart of this amazing universe, in which we are priviliged to participate, there is not division but unity!

References and Bibliography

Essential background reading for all who wish to study quantum theory is the collection of papers by J.S Bell, *Speakable and Unspeakable in Quantum Physics* (Cambridge, 1987). It contains his demonstration of Bell's theorem. The experiments that apparently support the non-local quantum predictions are reported in A. Aspect, P. Grangier and G. Roger, Phys. Rev. Letters, **49** (1982) 91. The best known remark of Newton regarding his belief in locality is contained in a letter to Bentley, 1692, reprinted in *Newton's Philosophy of Nature*, ed. H.S. Thaler (Hafner, New York, 1953). The basic physics behind the non-locality experiment discussed in Sect. 3 can be found in L. Hardy, *Phys. Rev. Letters,* **68** (1992), 2981. A possible way of retaining locality, compatible with present experiments, is given in my article "Has quantum non-locality been experimentally verified?" published in the Proceedings of the Cesena conference on Bell's Theorem and the Foundations of Modern Physics (1991), and also in E.J. Squires, Durham preprint, *A local hidden variable theory that, FAPP, agrees with experiment*, to be published in Physics Letters **A**.

I have written a much extended version of the discussion of this article in my book *Conscious Mind in the Physical World* (Adam Hilger, 1990). Related discussions from the point of view of a philosopher can be found in M. Lockwood, *Mind, Brain and the Quantum* (Blackwell, 1989), and from that of a judge in D. Hodgson, *The Mind Matters* (Oxford, 1991). The book by Penrose, *The Emperor's New Mind* (Oxford, 1989) contains some of the same ideas, and much more that is relevant.

Quantum Cosmology and the Emergence of a Classical World

Claus Kiefer

Institut für Theoretische Physik, Universität Freiburg,
Hermann-Herder-Str. 3,
D-79104 Freiburg, Germany

La pendule s'est arrêtée
Personne ne se bouge ...
Comme sur les images
Il n'y aura plus de nuit

Pierre Reverdy, from *La Réalité Immobile*

1. Why Quantum Cosmology?

Quantum cosmology is the application of quantum theory to the Universe as a whole. At first glance such an attempt seems surprising since one is used to apply quantum theory to microscopic systems. Why, then, does one wish to extrapolate it to the whole Universe? This extrapolation is based on the assumption that quantum theory is *universally valid*, in particular that there is no a priori classical world. The main motivation for this assumption comes from the kinematical non-locality (or non-separability) of quantum theory, i.e., from the fact that one cannot in general assign a wave function to a given system since it is not isolated but coupled to its natural environment, which is again coupled to another environment, and so forth. The extrapolation of this quantum entanglement thus leads inevitably to the concept of a wave function for the Universe. Many experiments, notably those which contradict Bell's inequalities, have impressively demonstrated this fundamental non-separability of quantum theory. As we will see in Sect. 3, it is this entanglement between many degrees of freedom which is also responsible for the emergence of a classical world. The importance of this effect seems to have been overlooked in the traditional discussion of quantum theory, and led to the belief in an independently existing classical world.

Since the dominant interaction in the cosmological realm is gravity, this extrapolation immediately has to address the problem of quantizing the gravitational field. In fact, since all other known interactions are successfully described by quantum field theories it seems unavoidable to quantize gravity, too, since the gravitational field is coupled to all other fields and it would appear strange, and probably even inconsistent, to have a drastically different framework for this one field. Many technical and conceptual difficulties have, however, as yet prevented the construction of a consistent and predictive theory of quantum gravity, some of which we will briefly describe in the course of this article.

Quantum cosmology is only meaningful as a physical theory if one can agree on how to interpret a universal quantum theory. The traditional "Copenhagen interpretation," for example, strictly denies the idea of a fundamental quantum world and assumes from the outset the existence of a classical world whose concepts have to be used to interpret the results of quantum measurements. Such an attitude appears to be ad hoc even in ordinary quantum mechanics since measurement apparata are built from atoms which are known to obey the laws of quantum theory. Although not consistent, such a hybrid description of Nature has been sanctioned by using purely verbal constructs like "complementarity." The conceptual inconsistency of the "Copenhagen interpretation" becomes even more evident in the framework of quantum cosmology, where the whole Universe including all observers and apparata *has to* be described in quantum terms from the very beginning. To quote Gell-Mann and Hartle (1990): "Quantum mechanics is best and most fundamentally understood in the framework of quantum cosmology."

But since it is impossible to prepare an ensemble of universes, can there be any fundamental meaning of a wave function of the Universe? The situation can be compared to ordinary quantum field theory where the concept of a vacuum state vector is used to derive properties of elementary particles but *not* to perform experiments with an ensemble of vacua. One might expect that in an analogous sense a wave function of the Universe, Ψ, determines structures of the Universe which can be checked by observations. One example is outlined in the last section of this article and is concerned with the direction of time. Another example is the claim (supported by heuristic considerations only) that Ψ is peaked around a vanishing cosmological constant (Baum, 1983) leading to the *prediction* that the cosmological constant *is* zero. Since such a wave function of the Universe is independent of any observation, one must attribute to it the status of reality.

As mentioned above, any attempt to understand quantum cosmology must focus on the construction of a quantum theory of gravity. One can distinguish several levels of the relationship between quantum theory and gravity which we briefly wish to describe. The most fundamental level should be described by a quantum theory of all interactions including gravity. A promising candidate in recent years has been the theory of superstrings which is constructed by using the assumption that the fundamental entities are one-dimensional objects instead of local quantum fields. Since many mathematical and conceptual problems in this framework have not yet been solved, no final agreement on the physical status of this "theory of everything" has been reached so far. A less ambitious level consists in the application of formal quantization rules to the general theory of relativity. Here one does not attempt to provide a unified theory of all interactions but tries to focus on the specific quantum aspects of the gravitational field. The scale where such aspects are expected to become relevant is found by setting the Compton wavelength associated with a mass equal to its Schwarzschild radius, and is called the Planck length:

$$l_{\text{Pl}} = \sqrt{\frac{\hbar G}{c^3}} \approx 10^{-33} \text{cm}. \tag{1}$$

One would expect that the level of quantum general relativity can be recovered from superstring theory in an appropriate low-energy limit. It might, however, also be the case that quantum general relativity *is* the fundamental theory, and impressive formal developments have taken place recently which support this possibility. We will comment on this below. In either case it is justified to study the implications of quantum general relativity, which is the very conservative framework for tying together quantum theory and general relativity. Both theories have up to now passed all tests in their respective realms.

On the next level we find the framework of quantum field theory for non-gravitational fields on a fixed classical background spacetime. How this level can be recovered from the more fundamental level of quantum general relativity, and in particular how the Schrödinger equation for non-gravitational fields can be derived from quantum gravity, is the subject of the third section. On this level one still has no experimental test but at least a definite prediction – the Hawking effect: black holes are not really black if quantum effects are taken into account, but radiate with a temperature

$$T = \frac{\hbar c^3}{8\pi G k_{\text{B}} M} = \frac{\hbar \kappa}{2\pi c k_{\text{B}}} \approx 10^{-6} \text{kelvin} \times \frac{M_{\text{sun}}}{M}, \tag{2}$$

where M is the mass of the black hole, k_{B} Boltzmann's constant, and κ the surface gravity of the hole. It is remarkable that all fundamental constants of nature appear in this formula. The significance of the relation (2) for a quantum theory of gravity might well turn out to be analogous to the significance of de Broglie's relation $p = \hbar k$ for the development of quantum mechanics (Zeh, 1992).

The experimental level of the relation between quantum theory and gravity is only reached at the very modest level of the Schrödinger equation with a Newtonian potential where impressive experiments using neutron interferometry have been performed.

What are the main obstacles in constructing a quantum theory of gravity? An important formal problem is the non-renormalizability of a perturbative expansion of quantum general relativity around a given background spacetime, i.e., the fact that an infinite number of parameters would have to be determined by experiment to absorb all the infinities of the theory. It is therefore not surprising that recent developments mainly focus on non-perturbative approaches. A serious obstacle for experiments in this field is the extreme smallness of the Planck length (1). One would have to extrapolate current particle accelerators to the dimensions of the Milky Way to be able to probe such a scale – a hopeless enterprise. It is thus hoped that a consistent theory of quantum gravity will reveal what the observable effects are. Such effects might well be found in a direction not imagined hitherto. One recent investigation, for example, claims that effects coming from the wave function of the Universe have already been observed by the COBE satellite (Salopek, 1993). Independently of the lack of experiments, there remain many conceptual problems which have to be addressed if gravity is to be quantized. Most of these problems are connected in one way or the other with the concept of time. We will discuss this issue in the next section but will first briefly outline the formal framework in which most of the recent investigations are made – the framework of *canonically* quantizing general relativity.

What does it mean to "quantize" a classical theory? The prominent role in quantum mechanics is played by the position and momentum operators (the "p's and the q's"). of a given system. The central property is their non-commutativity, which gives rise to the famous uncertainty relations. In the Schrödinger formulation of quantum mechanics one uses wave functions which are defined on configuration space, i.e., the space of all position coordinates (or, alternatively, on momentum space). "Canonically quantizing" a classical theory now means identifying the "p's and the q's" of the theory under consideration and imposing on them non-trivial commutation relations. In the case of the general theory of relativity this involves some preparatory steps since the p's can only be defined after an appropriate time parameter can be distinguished. Since the theory treats all space- and time-coordinates on an equal footing one has to formally rewrite the theory in such a manner that time-coordinates appear. This does not, of course, alter the physical content of the theory, which is invariant under general coordinate transformations.

Basically, the steps in canonically quantizing general relativity are the following. The first step consists in foliating spacetime into a family of spacelike three-dimensional hypersurfaces – one decomposes spacetime into space and time. The metric on these hypersurfaces, $h_{ab}(\mathbf{x})$, will play the role of the canonical variable (the q). All quantities are then decomposed into variables which live on such a hypersurface and variables which point in the fourth, timelike, dimension. It then turns out that the canonical momentum, $\pi^{ab}(\mathbf{x})$, is essentially given by the extrinsic curvature of a three-dimensional hypersurface, i.e., the quantity which describes the embedding of space into spacetime. Loosely speaking, one can say that intrinsic and extrinsic geometry are canonically conjugate to each other.

A major feature of general relativity is its invariance under arbitrary coordinate transformations. As a consequence one finds that there exist – at each space point – four constraints. Three of them generate coordinate transformations on the three-dimensional space and are analogous to Gauss' law in electrodynamics. The fourth, the so-called *Hamiltonian constraint*, plays a double role: Although a constraint, it generates the dynamics. Its explicit form is

$$\mathcal{H} \equiv \frac{16\pi G}{c^2} G_{abcd} \pi^{ab} \pi^{cd} - \frac{c^4}{16\pi G} \sqrt{h} R + \mathcal{H}_m = 0, \tag{3}$$

where $\sqrt{h}$ is the square root of the determinant of the three-metric, R is the curvature scalar on three-space, and $\mathcal{H}_m$ is the Hamiltonian density for non-gravitational fields. The coefficients G_{abcd} depend explicitly on the metric and themselves play the role of a metric in the space of all metrics. The existence of the constraint (3) is directly connected to the invariance of the theory under reparametrizations of time. It is therefore not surprising that it is quadratic in the momenta (the same happens, for example, in the case of the relativistic particle – invariance under reparametrizations of the world line parameter leads to the constraint $p^2 + m^2 = 0$). The presence of the invariance under coordinate transformations and its associated constraints lies at the heart of the quantization problem. To quote Pauli (1955):

> Es scheint mir ..., daß nicht so sehr die Linearität oder Nichtlinearität Kern der Sache ist, sondern eben der Umstand, daß hier eine allgemeinere Gruppe als die Lorentzgruppe vorhanden ist ...[1]

Quantization now proceeds, at least formally, by elevating the metric and its momentum to the status of operators and imposing the commutation relations

$$[h_{ab}(\mathbf{x}), \pi^{cd}(\mathbf{y})] = i\hbar \delta^c_{(a} \delta^d_{b)} \delta(\mathbf{x}, \mathbf{y}) \tag{4}$$

in full analogy to the commutation relation $[q, p] = i\hbar$ in quantum mechanics. One specific realization of (4) is provided by the substitution (in analogy to substituting $p \to \frac{\hbar}{i} \frac{d}{dq}$)

$$\pi^{ab} \to \frac{\hbar}{i} \frac{\delta}{\delta h_{ab}}. \tag{5}$$

The classical constraint (3) is then formally implemented in the quantum theory by inserting (4) and (5) into (3) and applying it on wave functionals Ψ which depend – apart from non-gravitational fields denoted by ϕ – on the three-metric, i.e.,

$$\mathcal{H}\Psi[h_{ab}(\mathbf{x}), \phi(\mathbf{x})] = \left(-\frac{16\pi\hbar^2 G}{c^2} G_{abcd} \frac{\delta^2}{\delta h_{ab}\delta h_{cd}} - \frac{c^4}{16\pi G} \sqrt{h} R + \mathcal{H}_m \right) \Psi = 0. \tag{6}$$

This equation is called the *Wheeler-DeWitt equation* in honour of the pioneering work of DeWitt (1967) and Wheeler (1968). It has the form of a zero-energy

---

[1] It seems to me ... that it is not so much the linearity or non-linearity which forms the heart of the matter, but the very fact that here a more general group than the Lorentz group is present ...

Schrödinger equation. There are of course many technical problems such as factor ordering or regularization which we will not be able to address in this article (see, e.g., Kuchař, 1992).

The quantization of the remaining three constraints leads to the condition that this wave functional actually does not change under a coordinate transformation of the three-metric but is a function of the *geometry* only. The configuration space is thus the space of all three-geometries and is called *superspace*. An important physical consequence of the commutation rules (4) is the "uncertainty" between the three-dimensional space and its embedding in the fourth dimension: The concept of spacetime is a classical concept only with no fundamental meaning in quantum theory. This is fully analogous to the concept of a particle trajectory which has no fundamental meaning in quantum mechanics due to the uncertainty between position and momentum. When applied to cosmology, the wave functional Ψ in (6) is the desired "wave function of the Universe" which was mentioned at the beginning of this section.

An important development has been the discovery of appropriate variables which enables one to find exact formal solutions of equation (6) in the absence of matter (Ashtekar, 1991). This became possible since the complicated potential term of (6) disappears when the equation is written in terms of the new variables, which have a strong similarity to Yang-Mills variables. The solutions can be classified in terms of loops and knots and exhibit an interesting structure of space, of which maybe the most important is the existence of a minimal length of the order of the Planck length (1). Consequently, smaller scales do not have any operational meaning.

We have not yet addressed the issue of boundary conditions to be imposed on the Wheeler-DeWitt equation. In contrast to systems in the laboratory, they are not at our disposal. In fact, the question of boundary conditions has been of great interest in recent years, basically because of the *no boundary proposal* by Hartle and Hawking (1983). These two authors express the wave functional as a formal path integral where the sum is over euclidean (instead of lorentzian) geometries. The no boundary proposal then consists in the condition that one performs a sum over compact manifolds with one boundary only – the boundary which is given by the considered universe. The lack of a second boundary (for example at a small size of the universe) saves one from the need to find appropriate boundary conditions there. Unfortunately, these path integrals can only be evaluated in a semiclassical approximation and, moreover, do not lead to a unique wave function. We will return to the question of boundary conditions in the last section in connection with a discussion of the arrow of time.

Since the full equation (6) is in general difficult to handle, most applications have been performed in a restricted framework where only finitely many degrees of freedom are quantized. A typical example in the cosmological context is the quantization of the scale factor a of a Friedmann Universe. If in addition a (conformally coupled) scalar field ϕ is taken into account to simulate the matter content of the Universe we find instead of (6) the much simpler equation

$$H\psi(a,\phi) = \left(\frac{\partial^2}{\partial a^2} - \frac{\partial^2}{\partial \phi^2} - a^2 + \phi^2 \right) \psi(a,\phi) = 0. \tag{7}$$

This equation has the form of an indefinite harmonic oscillator and can serve as a useful tool in studying conceptual questions in quantum cosmology (see the following sections).

The canonical quantization scheme outlined in this section still contains the structure of a differentiable three-dimensional manifold. There exist more ambitious approaches to quantum gravity which give up even this structure and start with a quantization of topology, see, e.g., Isham, Kubyshin and Renteln (1990), who construct wave functions which live on a family of topologies. We will not, however, follow these approaches any further.

2. Time in Quantum Gravity

In the preface to Max Jammer's book *The Problem of Space*, Einstein writes

> Es hat schweren Ringens bedurft, um zu dem für die theoretische Entwicklung unentbehrlichen Begriff des selbständigen und absoluten Raumes zu gelangen. Und es hat nicht geringerer Anstrengung bedurft, um diesen Begriff nachträglich wieder zu überwinden – ein Prozeß, der wahrscheinlich noch keineswegs beendet ist.[2]

Although Einstein refers in this quotation to space only, the same can be said about time. Absolute time, as well as absolute space, was an indispensable ingredient in Newton's theory of motion and more powerful in the development of dynamics than, for example, the notion of relative time which was put forward by Leibniz. Time kept its absolute status even in non-relativistic quantum mechanics – the t in Schrödinger's equation is still an external parameter and not turned into a quantum operator. Although the causal structure of spacetime has drastically changed with the advent of the special theory of relativity by dropping the notion of absolute simultaneity, spacetime is there still understood as a *non*-dynamical entity which provides an arena for the laws of physics but does not take part in the play. There is hence still no quantum operator for time in relativistic quantum field theory, although one now has the possibility to choose any spacelike hypersurface as a time parameter. These spacelike hypersurfaces can be deformed independently at each space point, which gives rise to a local or "many-fingered" time $\tau(\mathbf{x})$ instead of one single t. This local time appears also in the field-theoretic Schrödinger

---

[2] It was a hard struggle to gain the concept of independent and absolute space which is indispensible for the theoretical development. And it was no lesser effort to overcome this concept later on, a process which probably has not yet come to an end.

equation, which is an equation for wave *functionals* depending on fields $\phi(\mathbf{x})$ (which play the role of the "*q*'s" in quantum mechanics). This equation reads

$$i\hbar \frac{\delta\psi[\phi(\mathbf{x})]}{\delta\tau(\mathbf{x})} = \mathcal{H}_m\psi[\phi(\mathbf{x})], \qquad (8)$$

where $\mathcal{H}_m$ is the Hamiltonian density connected with $\phi(\mathbf{x})$.

Spacetime becomes a *dynamical* object, in analogy to a particle trajectory, only in the general theory of relativity where the spacetime metric describes the gravitational field, which is subject to Einstein's field equations. The quantization of gravity therefore unavoidably has to address the quantization of time. Why would one expect any problems to occur in this step? One has to recall that the presence of an external time is an essential ingredient of quantum mechanics – matrix elements are calculated at a given time, and measurements are performed at a given time. Moreover, probabilities are preserved *in* time. The *absence* of any time parameter is a fundamental property of the Wheeler-DeWitt equation (6). This is not surprising since, as we have argued above, the concept of spacetime has no fundamental meaning in quantum gravity. The question therefore arises: Can one introduce a viable concept of time on the fundamental level of quantum gravity itself or only in a semiclassical approximation? A critical investigation into the concept of time may lead to fruitful insights into the structure of the desired theory. To quote Einstein again:

> ... und doch ist es im Interesse der Wissenschaft nötig, daß immer wieder an diesen fundamentalen Begriffen Kritik geübt wird, damit wir nicht unwissentlich von ihnen beherrscht werden.[3]

The emergence of a semiclassical time from quantum gravity will be discussed in the next section. Here we focus on the level of equation (6) itself. An important property of the Wheeler-DeWitt equation is its hyperbolic nature, i.e., its behaviour as a wave equation. This can be recognized from the coefficients G_{abcd} by treating them – at each space point separately – as the elements of a symmetric 6×6 matrix which after diagonalization has the signature $(-, +, +, +, +, +)$. It is important to note that there remains *one* global minus sign after gauge degrees of freedom have been eliminated. This minus sign is basically given by the size of the Universe, which may thus be considered as an *intrinsic time* variable. Ψ has now to be interpreted as a probability amplitude *for* time, but not *in* time. All other degrees of freedom, including physical clocks, are correlated with intrinsic time. The hyperbolic nature of (6) allows the formulation of a Cauchy problem with respect to this time. In spite of its static appearance, the Wheeler-DeWitt equation describes an intrinsic dynamics! This has drastic consequences for the behaviour of wave

---

[3] ... and yet it is necessary, in the interest of science, to call these fundamental concepts again and again into question so that we are not governed by them without realising it.

packets in the case of a recollapsing universe (Zeh, 1988). In the classical theory, the recollapsing leg of the history of the universe can be considered as the deterministic successor of the expanding leg. This is no longer true in quantum cosmology! The wave equation (6) describes dynamics with respect to intrinsic time, which in simple models like (7) is the radius of the universe. The expanding and recollapsing legs of a wave packet concentrated near the classical trajectory in configuration space can thus not be distinguished intrinsically. With respect to the Cauchy problem the returning wave packet must be present initially.

This can already be seen in the simple model (7). To construct a wave tube which follows the collapsing trajectory, one must use solutions to (7) which fall off for large values of a and ϕ. This forces one to use the normalisable harmonic oscillator eigenfunctions in a wave packet solution which thus reads

$$\psi(a,\phi) = \sum_n A_n \frac{H_n(a)H_n(\phi)}{2^n n!} \exp\left(-\frac{1}{2}a^2 - \frac{1}{2}\phi^2\right), \qquad (9)$$

where A_n are coefficients which are peaked around some $n = \bar{n}$, and H_n are the Hermite polynomials. The solution (9) automatically describes *two* packets at $a = 0$ (see Fig. 1). While in the simple oscillator model (7) there is no dispersion of the wave packet, this is no longer true in more realistic examples like the case of a massive scalar field in a Friedmann universe (Kiefer, 1988; Zeh, 1992). The demand for the wave packet to go to zero at large radii (otherwise it cannot correspond to a recollapsing universe) unavoidably leads to a *smearing* of the packet in regions close to the classical turning point. This demonstrates that a WKB approximation cannot hold in the whole region of an expanding and recollapsing trajectory in configuration space. This has important consequences for the discussion of the arrow of time (see the last section).

The above considerations are at present only of a heuristic nature. It is still unclear what the Hilbert space structure of quantum theory is, if it is necessary at all. Imposing the Schrödinger inner product onto the solutions of (6) in general leads to a diverging result if *all* variables are integrated over. Recalling the hyperbolic structure of this equation, it would seem to be more appropriate to choose a Klein-Gordon inner product like in relativistic quantum mechanics and to integrate over "spacelike" hypersurfaces $a = constant$. Unlike relativistic quantum mechanics ,however, it is not here possible to make a decomposition into positive and negative frequencies (see Kuchař, 1992). This has prompted some authors to invoke a third quantization of the theory, i.e., to elevate the wave functional Ψ to an operator in some "new" Hilbert space. However, we will not follow these approaches any further. We also mention that there are attempts which try to *first* solve the classical constraint (3) and only *then* make the transition to quantum theory. This creates its own problems, and we refer to the excellent review articles by Isham (1992) and Kuchař (1992) for details.

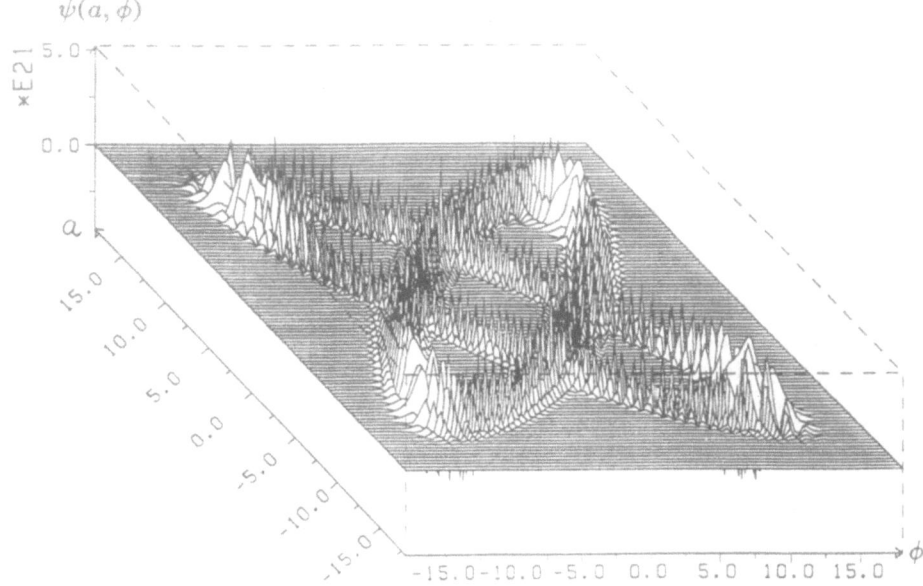

$\psi(a, \phi)$

Fig. 1. Wave packet solution of the Wheeler-DeWitt equation (7)

3. Decoherence and the Recovery of the Schrödinger Equation

We now address the issue of how the level of quantum field theory in a classical spacetime background can be recovered from quantum gravity where there is no spacetime. This will basically involve two steps. The first step is the derivation of the Schrödinger equation (8) from the Wheeler-DeWitt equation (6). The second step is to understand the unobservability of non-classical states for the gravitational field. A detailed review of these steps can be found, e.g., in Kiefer (1993b).

The basic observation which goes into the development of the first step is the fact that the length scale contained in (6), the Planck length (1), is much smaller than any relevant scale of non-gravitational physics. This enables one to make a formal expansion of the wave functional in (6) in powers of the Planck length (or, equivalently, the gravitational constant). If there were no non-gravitational fields in (6), an expansion with respect to G would be fully equivalent to an expansion in powers of $\hbar$ which is the usual semiclassical ("WKB") expansion, since both constants would appear in the combination $G\hbar$ only. As far as gravity is concerned, the present expansion scheme thus *is* a WKB expansion. This is no longer true for other fields, in which case the situation is analogous to a Born-Oppenheimer approximation in molecular physics: the gravitational part in (6) corresponds to the heavy nuclei whose kinetic terms are neglected in a first approximation while the remaining part corresponds to the light electrons. To highest order, the wave functional depends only on the gravitational field,

$$\Psi_0 = C[h_{ab}] \exp\left(\frac{ic^2}{32\pi G\hbar} S_0[h_{ab}]\right),\tag{10}$$

where S_0 obeys the Hamilton-Jacobi equation for gravity. This equation is equivalent to all of Einstein's field equations and describes a classical gravitational background in the sense that one can assign classical "trajectories" to it. Each trajectory, which represents a whole spacetime, runs orthogonally to hyperspaces S_0 = constant in configuration space. Formally, this is the same as the recovery of geometrical optics from Maxwell's equations.

In the next order of approximation the wave functional assumes the form

$$\Psi_1 = \Psi_0 \, \chi[h_{ab}, \phi],\tag{11}$$

where the wave functional χ also depends on non-gravitational fields and obeys the equation (Banks, 1985)

$$i\hbar G_{abcd}\frac{\delta S_0}{\delta h_{ab}}\frac{\delta \chi}{\delta h_{cd}} \equiv i\hbar\frac{\delta \chi}{\delta \tau(\mathbf{x})} = \mathcal{H}_m\chi.\tag{12}$$

This is nothing else but the functional Schrödinger equation (8) for non-gravitational fields propagating on *one* of the classical spacetimes described by S_0. This spacetime is parametrized by the "many-fingered time" $\tau(\mathbf{x})$ appearing in (12).

As Julian Barbour (1992) emphasized, this WKB-time corresponds exactly to the notion of *ephemeris time* used by astronomers. Ephemeris time is the extraction of time, in retrospect, from actual observations of celestial bodies, see for example Clemence (1957). The semiclassical time in (12) is thus defined by the actual motion of bodies in the real world. It is amazing that this exactly corresponds to the concept of time used by the ancient Greeks who defined time by the motion of the celestial bodies (already Plato used the term ephemeris time). It is clear that there can be no emergence of a semiclassical time for flat Minkowski space, which demonstrates the absence of any concept of absolute time. It is the three-dimensional geometry which carries information about time, see Baierlein, Sharp and Wheeler (1962). To determine ephemeris time, eventually all available information about motion in the Universe has to be taken into account. An impressive example is the case of the binary pulsar PSR 1913+16 . Using general relativity, ephemeris time can only be consistently extracted from the orbital motion if the gravitational pull of the whole Galaxy on the pulsar is taken into account, as was demonstrated by Damour and Taylor (1991).

If one proceeds with the above approximation scheme to the next order (Kiefer and Singh, 1991), one can derive correction terms to the Schrödinger equation (12) which are proportional to the gravitational constant. In addition one finds a back reaction of the non-gravitational fields onto the gravitational background which modifies the definition of semiclassical time. One can calculate concrete results from these correction terms, such as the quantum gravitational correction to the trace anomaly in De Sitter space (Kiefer, 1993b).

Is the recovery of the Schrödinger equation in a classical spacetime sufficient for the understanding of the classical behaviour of the spacetime geometry in our world? The answer must be *no* since one still has to focus on the issue of superpositions of different "classical" states of the gravitational field. This problem is even of direct relevance in the above derivation: if one takes a superposition of two semiclassical states, for example the state (10) and its complex conjugate, one *cannot* recover the Schrödinger equation (Barbour, 1993). This equation follows only if a special, *complex*, state like (10) is taken as the starting point. As one recognizes immediately, this issue is directly connected with the emergence of the i and the use of complex wave functions in ordinary quantum theory – the i in the Schrödinger equation is taken directly from the state (10) of the gravitational field. How, then, can one justify the use of such a special state? A possible answer is provided by the mechanism of decoherence (Kiefer, 1993a). The key ingredient is the quantum entanglement of the wave function of the Universe, i.e., the existence of correlations between a large number of degrees of freedom, which we discussed in the first section. Since only very few degrees of freedom in this wave function are accessible to observations, the relevant object is *not* the full wave function but the reduced density matrix which is obtained by tracing out all irrelevant (unobservable) degrees of freedom. If the only relevant degree of freedom were the scale factor, this density matrix would read

$$\rho(a, a') = \mathrm{Tr}_\phi \Psi^*[a', \phi]\Psi[a, \phi], \tag{13}$$

where ϕ stands for the irrelevant degrees of freedom. These can be gravitational degrees of freedom like gravitational waves as well as, for example, matter density perturbations. They "measure" the gravitational background and force it to become classical (Zeh, 1986; Kiefer, 1987). In this way "quasiclassical domains" emerge from the fundamental quantum world (Gell-Mann and Hartle, 1990). When applied to simple models, the tracing out of such irrelevant degrees of freedom suppresses interference terms between different WKB components of the form (10). A conformally coupled scalar field in a Friedmann universe, for example, leads to a suppression factor of the interference between (10) and its complex conjugate given by

$$\exp\left(-\frac{\pi m H_0^2 a^3}{128}\right), \tag{14}$$

where m, H_0, and a are, respectively, the mass of the scalar field, the Hubble parameter, and the scale factor (Kiefer, 1992). This is very tiny, except for small radii of the Universe and near the region of the classical turning point where the Hubble parameter vanishes. Quantum gravity itself thus contains the seeds for the emergence of a classical geometry, but also describes its limit.

4. The Direction of Time

It is an obvious fact that most phenomena in Nature distinguish a direction of time (see, for example, Zeh, 1992): Electromagnetic waves are observed in their retarded form only, where the fields causally follow from their sources. The increase of entropy, as expressed in the second law of thermodynamics, also defines a time direction. This is directly connected with the psychological arrow of time – we remember the past but not the future. In quantum mechanics it is the irreversible measurement process and in cosmology the expansion of the Universe, as well as the local growing of inhomogeneities, which determine a direction of time.

And yet, the fundamental laws of physics are invariant under time reversal (the only exception being the small CP violation in weak interactions). How, then, can one understand that most phenomena distinguish a direction of time? The answer lies in the possibility of *very special* boundary conditions such as an initial condition of low entropy.

One assumes now that the occurrence of such boundary conditions can be understood within the dynamical laws of physics and that it makes sense to search for a common root of the various arrows of time – the *master arrow of time*. Such an assumption transcends the Newtonian separation into laws and boundary conditions by also seeking physical explanations for the latter.

Where lies the key to the understanding of the irreversibility of time? As in particular Penrose (1979) has convincingly emphasized, it is primarily the high unoccupied entropy capacity of the gravitational field which allows for the emergence of structure far from thermodynamical equilibrium. Whereas the non-gravitational part of the entropy reaches its maximum for a homogeneous state, the opposite is true for the gravitational part which tries to develop a highly clumpy state. It is therefore a cosmological problem to justify the presence of an initial state of very low gravitational entropy, i.e. a very homogeneous state. This has provoked Penrose to formulate his *Weyl tensor hypothesis* that the Weyl tensor vanishes at singularities in the past but not at those in the future. The Weyl tensor is that part of the Riemann tensor which is not fixed by the field equations (in which only the Ricci tensor enters) but by the boundary conditions only. It describes the degrees of freedom of the gravitational field. Since it vanishes exactly for a homogeneous and isotropic Friedmann Universe, it can be taken as a heuristic measure for inhomogeneity and, therefore, for gravitational entropy.

The arrow of time can of course only be explained by the Weyl tensor hypothesis if it can be derived from some fundamental theory. As we have argued in the previous sections, the fundamental framework to address such questions is quantum cosmology. In the configuration space for the wave function of the Universe there is no intrinsic distinction between Big Bang and Big Crunch since both just correspond to regions of low scale factor a. The desired boundary condition is then a boundary condition for the wave function at small scales. It has therefore been suggested (Zeh, 1993) to impose the boundary condition that for small scales the wave function depends only on a but *not* on any other degrees of freedom. These degrees of freedom emerge only with increasing a when they are in their ground state, as can be seen from a discussion of the Wheeler-DeWitt equation (Conradi, 1992). A similar behaviour was also derived from the no boundary condition (Hawking, 1985). At least heuristically, this implements a low gravitational entropy at small scales. Complexity, and therefore entropy, increases with increasing scale factor. Since more and more degrees of freedom come into play, decoherence also increases and the Universe becomes more and more classical. The thermodynamical arrow of time is thus inextricably tied to the cosmological arrow of time.

Such a boundary condition has important physical consequences. Since the thermodynamical arrow of time is correlated with the scale factor but not with any classical trajectory (which is absent in quantum cosmology), this would mean that in the case of a closed universe time would formally reverse its direction near the turning point. We call such a reverse formal since no information gathering system could survive this reversal. All observers in any branch of the wave function would consider their universe as expanding. This boundary condition would also have drastic consequences for the behaviour of black holes (Zeh, 1993) since they would formally become white holes during the recontraction phase, and the formation of a horizon would be prevented. The whole quantum universe would be singularity-free in this case.

These considerations are of course still speculative. However, they are only based on the established formal framework of quantum theory and cosmology. They demonstrate that quantum cosmology, if taken seriously, can yield a picture of the Universe which drastically modifies the classical Big Bang model.

Acknowledgement

I am grateful to H.-Dieter Zeh for many stimulating discussions and a critical reading of this manuscript. I also thank Tejinder Singh for his comments on this manuscript.

References

A. Ashtekar (1991): *Lectures on non-perturbative canonical gravity.* World Scientific, Singapore.

R. G. Baierlein, D. H. Sharp, and J. A. Wheeler (1962): Three-Dimensional Geometry as Carrier of Information about Time. Phys. Rev. **126**, 1864.

T. Banks (1985): TCP, Quantum Gravity, the Cosmological Constant, and all that .... Nucl. Phys. **B249**, 332.

J. B. Barbour (1992): Personal communication.

J. B. Barbour (1993): Time and complex numbers in canonical quantum gravity. Phys. Rev. D **47**, 5422.

E. Baum (1983): Zero cosmological constant from minimum action. Phys. Lett. **B133**, 185.

G. M. Clemence (1957): Astronomical Time. Rev. Mod. Phys. **29**, 2.

H. D. Conradi (1992): Initial state in quantum cosmology. Phys. Rev. D **46**, 612.

T. Damour and J. H. Taylor (1991): On the orbital period change of the binary pulsar PSR 1913+16. Astrophys. Journal **366**, 501.

B. S. DeWitt (1967): Quantum Theory of Gravity I. The Canonical Theory. Phys. Rev. **160**, 1113.

M. Gell-Mann and J. B. Hartle (1990): Quantum Mechanics in the Light of Quantum Cosmology. In *Complexity, Entropy and the Physics of Information*, ed. by W. H. Zurek. Addison Wesley.

J. B. Hartle and S. W. Hawking (1983): Wave Function of the Universe. Phys. Rev. D **28**, 2960.

S. W. Hawking (1985): Arrow of Time in Cosmology. Phys. Rev. D **32**, 2489.

C. J. Isham (1992): Canonical Quantum Gravity and the Problem of Time. Lectures presented at the NATO Advanced Study Institute *Recent problems in Mathematical Physics*, Salamanca, June 15-27, 1992.

C. J. Isham, Y. Kubyshin, and R. Renteln (1990): Quantum norm theory and the quantisation of metric topology. Class. Quantum Grav. **7**, 1053.

C. Kiefer (1987): Continuous measurement of mini-superspace variables by higher multipoles. Class. Quantum Grav. **4**, 1369.

C. Kiefer (1988): Wave packets in minisuperspace. Phys. Rev. D **38**, 1761.

C. Kiefer (1992): Decoherence in quantum electrodynamics and quantum gravity. Phys. Rev. D **46**, 1658.

C. Kiefer (1993a): Topology, decoherence, and semiclassical gravity. Phys. Rev. D **47**, 5414.

C. Kiefer (1993b): The Semiclassical Approximation to Quantum Gravity. To appear in *The Canonical Formalism in Classical and Quantum General Relativity.* Springer, Berlin.

C. Kiefer and T. P. Singh (1991): Quantum Gravitational Corrections to the Functional Schrödinger Equation. Phys. Rev. D **44**, 1067.

K. V. Kuchař (1992): Time and Interpretations of Quantum Gravity. In *Proceedings of the 4th Canadian Conference on General Relativity and Relativistic Astrophysics*, ed. by G. Kunstatter, D. Vincent and J. Williams. World Scientific, Singapore.

W. Pauli (1955): Schlußwort. Helv. Phys. Acta Suppl. **4**, 266.

R. Penrose (1979): Singularities and Time-Asymmetry. In *General Relativity*, ed. by S. W. Hawking and W. Israel. Cambridge University Press, Cambridge.

D. S. Salopek (1993): Searching for Quantum Gravity Effects in Cosmological Data. To appear in the *Proceedings of Les Journées Relativistes*, Brussels, April 5-7, 1993. World Scientific, Singapore.

J. A. Wheeler (1968): Superspace and the nature of quantum geometrodynamics. In *Battelle rencontres*, ed. by C. M. DeWitt and J. A. Wheeler. Benjamin.

H. D. Zeh (1986): Emergence of Classical Time from a Universal Wave Function. Phys. Lett. **A116**, 9.

H. D. Zeh (1988): Time in Quantum Gravity. Phys. Lett. **A126**, 311.

H. D. Zeh (1992): *The Physical Basis of The Direction of Time*. Springer, Berlin.

H. D. Zeh (1993): Time (a-)symmetry in a recollapsing universe. In *Physical Origin of Time Asymmetry*, ed. by J. J. Halliwell, J. Perez-Mercader and W. H. Zurek. Cambridge University Press, Cambridge.

On the Origin of Structure in the Universe

Julian B. Barbour

College Farm, South Newington, Banbury, Oxon, OX15 4JG, England

1. Introduction

The most obvious thing about the universe in which we find ourselves is its structure. Indeed, according to the basic philosophical viewpoint adopted by thinkers such as Berkeley, Leibniz and Mach, it would not really be possible to speak of the universe at all if it did not exhibit differentiating structure and qualities. Leibniz's monadology equates existence with variety: monads are defined by and simultaneously distinguished from other monads by virtue of their attributes and nothing else. You cannot remove the attributes of a thing and leave some mysterious "thisness" (*haeccity*). Remove the attributes and nothing is left.

The approach of modern science to the fact of the structured universe is confused and ambivalent. It certainly does not make a frontal attack on the problem of structure. The reason for this is rather interesting and possibly instructive. Before the scientific revolution, the instinctive reaction of thinkers to the existence of perceived structure was to find a direct reason for that structure. This is reflected above all in the Pythagorean notion of the well-ordered cosmos. The cosmos has the structure it does because that is the best structure it could have. In fact, that is what the word *cosmos* really means – primarily order, but also *decoration, embellishment,* or *dress (cosmetic* has the same origin). Kepler and Galileo were no less entranced by the beauty of the world than Pythagoras and formulated their ideas in the overall framework of the notion of the well-ordered cosmos. Indeed, both saw their work as contributing significantly to the articulation of the idea – Kepler (1596) with his Platonic solids, while Galileo (1632) regarded free fall under gravity as "returning a misplaced object [a stone raised from the ground] to its proper place."

However, both looked at the world so intently that they actually identified aspects of motion (precise laws of planetary motion and simple laws of falling bodies and projectiles) that relatively soon led to the complete over-throw of such a notion of cosmos. The laws of the new physics were found to determine, not the actual structure of the universe, but the way in which structure changes from instant to instant. No explanation is provided for the presently observed structure; it is simply attributed to an initial structure that was never *fashioned* by the laws of nature but merely continually *refashioned* thereafter. The initial and boundary conditions for our universe lie outside the purview of science. But all the structure we observe around us must ultimately be traced back to those mysterious initial and boundary conditions.

It is true that there are a few cases in which science attempts to under-stand structure in terms of dynamics, but dynamical laws by themselves are never sufficient. I shall mention just two examples, which seem to be the best substantiated and understood, and not discuss, for example, the speculative idea that the 'seeds' for nucleation of the galaxies were quantum fluctuations in the early universe. The first example is dynamically self-organizing struc-tures, on which much work has been done, above all by the Brussels school led by Ilya Prigogine (1980). However, a key requirement here is the existence of strong disequilibrium as an initial (and also ambient) condition. In such an approach, the structure of the universe is ultimately explained by the action of the dynamical laws on a most unlikely (from a purely statistical point of view) initial condition. This is, of course, intimately related to the famous problem of the origin of the second law of thermodynamics.

Structure arises (formally at least) in a totally different way in quantum mechanics. Schrödinger's *time-independent* wave equation for energy eigen-states determines states in which each possible configuration of the complete system under consideration is present with some definite probability. Cer-tain configurations of the system turn out to be vastly more probable than others; in particular, the most probable structures can sometimes exhibit an extraordinarily high degree of correlation between the parts, as, for example, in a crystal lattice. Since such are the structures one is most likely to find as a result of an observation (in accordance with the standard interpreta-tion of quantum mechanics), the time-independent Schrödinger equation can be said to generate structure. However, a crucial part of the story has not yet been mentioned: the solution of the Schrödinger equation must satisfy certain boundary conditions. Moreover, even when the boundary conditions have been fixed, there can, in principle, still be many different eigensolutions, corresponding to different energy eigenvalues (and even, in the case of degen-eracy, to one and the same eigenvalue). So here too it is not possible to create structure out of a dynamical equation by itself.

I should like to suggest that the concentration of science on dynamical laws could in the end turn out to have been the result of an historical accident, and that it may be time to return to the more primal quest for an explanation of structure in its own right.

I should therefore like to consider a very different way in which structure might be created. Since FESt, the organizer of this workshop, has a long-term interest in the philosophy of Leibniz, I hope that indulgence will be shown to ideas that might otherwise be regarded as grossly speculative – more appropriate to "armchair" physics than the real stuff. However, as, by his own account, Descartes hatched out the philosophical scheme that was to become the framework of Western science *in* a stove somewhere in Bavaria [quoted by Russell (1945): "Descartes *says* it was a stove"], I do not see why we should not have a shot at an alternative scheme – for it is the Cartesian scheme that needs to be overthrown – from an armchair in Heidelberg.

What Leibniz objected to above all in the amalgam of Cartesian and New-tonian ideas that was taking shape at the end of his life as the basic framework of modern science was the scant attention paid to the *individuation of objects*. Leibniz believed that true individuation by genuine attributes was the *sine qua non* of existence. Hence his objection to Newtonian absolute space, all of whose points were assumed to be absolutely indistinguishable among them-selves.

Leibniz developed what to many is a quite fantastic philosophy but to me is the one radical alternative to Cartesian-Newtonian materialism ever put forward that possesses enough definiteness to be cast in mathematical form – and hence serve as a framework for natural science.

According to Leibniz, the entire world consists of nothing but distinct individuals. The sole essence of these individuals is to have perceptions (not all of which they are distinctly aware of). Thus far Leibniz's philosophy is very similar to Berkeley's idealism (according to Berkeley nothing exists except perceiving souls and ideas – perceptions – which God causes to appear to them). What distinguishes it from Berkelian idealism is the most radical (and least well explained by Leibniz) element in the *Monadology* – the assertion that the perceptions of any one monad are nothing more and nothing less than the *relations* it bears to all other monads.

The complete set of monads thus exists by virtue of self-mirroring of each other. A monadological world is a perfectly bootstrapped world: tugged by itself into existence out of the mire of nothing somewhat after the manner that Münchhausen got himself out of the bog!

A few years ago my collaborator Lee Smolin (1988) formulated what I believe is the first mathematical model that gives genuine expression to many (I will not claim all) of the ideas that underline the *Monadology*. I shall present the basic idea of his scheme from a distinctly Leibnizian viewpoint that takes seriously the idea that the basic entities of the world, the monads, are indeed sentient beings. However, the mathematical scheme is independent of this interpretation. The basic entities of Smolin's scheme may perfectly well be conceived as material particles or fundamental events.

2. The Model

The model is based on the mathematical concept of a *graph*. In the simplest form, used here, a graph consists of just two types of entities: *vertices* and *lines*. I shall assume that the number N of vertices is fixed; it may be arbitrarily large but must be finite. There is one and only one relationship that can hold between two different vertices – either they are joined by a single line or they are not. In general graph theory, vertices can be joined by any number of lines, which may also have a definite sense, i.e., 'point' in one direction (from vertex A to vertex B). However, to simplify the model, I shall assume that the lines are undirected and that any two vertices are joined either by a single line or not at all. I shall call such a graph an *allowed graph*; such graphs can be represented as in Fig. 1. It is important that the position of the vertices and the lengths of the lines joining them in the pictorial representation have no significance. All that counts is whether or not any two vertices are connected or not. A graph is said to be *connected* if one can pass from any vertex of the graph to any other by a path formed by lines of the graph. All graphs considered in this paper will be connected.

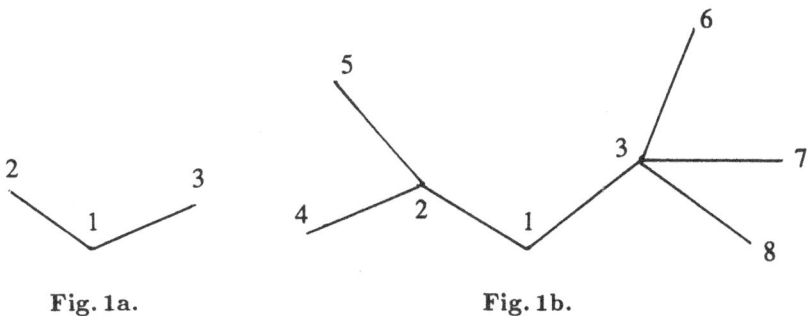

Fig. 1a.   Fig. 1b.

In many mathematical applications, graphs are used to represent certain salient relations within structures from which features inessential for the problem under consideration have been abstracted. Thus, in the very first application of graph theory, by Euler, the vertices represented places in Königsberg and the lines represented bridges over rivers. Euler was solely concerned with the problem of whether all the places could be reached in a continuous walk that did not cross any of the bridges more than once. In such problems, the lines and vertices can be defined and identified by attributes that are not represented in the graph. Thus, in general graph theory many vertices may have no connections at all by lines to other vertices, but one may still suppose the vertices distinct and as truly existing. They are identified by *extrinsic* denominations.

However, for the purposes of modelling the *Monadology*, I shall insist that all denominations are purely *intrinsic*. Thus, I am not allowed simply to point at what is shown in Fig. 1a as vertex 1 and say : "That is vertex 1." Instead, I must identify the vertex solely by means of the graph-theoretical elements that are present. For example, in Fig. 1a vertex 1 can be unambiguously distinguished from the other two by saying that it is joined to *two* other vertices. Note that the other two vertices are completely indistinguishable in graph-theoretical terms. I shall call such a graph *non-Leibnizian*, since those two vertices are an affront to Leibniz's principle of the identity of indiscernibles. In contrast, a graph will be *Leibnizian* if all of its vertices are unambiguously distinguishable by means of the intrinsic attributes allowed in this scheme. For graphs with a small number of vertices, it is quite difficult to find any that are Leibnizian. However, with increasing number of vertices, the proportion of all connected graphs with a given number of vertices that are non-Leibnizian decreases quite rapidly as a fraction of the total number of such graphs.

Let me now define the *r-step view* of any given vertex in a graph of the kind we are considering. It is simply the subgraph obtained from the complete graph by starting at the vertex under consideration and keeping all of the vertices and lines that can be reached (respectively traversed) in not more than r steps from the chosen vertex. For example, the 1-step view of vertex 1 in Fig. 1b is the complete graph in Fig. 1a. The 2-step view of vertex 1 in Fig. 1b is already the complete graph, but for vertex 8 the complete graph is recovered only with the 4-step view. It is obvious that the largest possible value that r can ever reach before the complete graph is included in the r-step view of any vertex is $N - 1$, though for most graphs this will happen long before such a value is reached.

The concept of r-step views can be used to distinguish vertices. If there are no intrinsic differences between the r-step views of two vertices, then we can say that they are r-step indistinguishable, or *r-indifferent*. But if there are intrinsic differences between the views, then they are *r-distinct*.

Now consider a particular vertex in a given Leibnizian graph of N vertices. We can define its *vertex indifference* as the minimal value of r at which it becomes r-distinct from *every* other vertex in the graph. For the complete graph, there are N such vertex indifferences. If we add them up, we obtain a positive number, which must be at least N, that we can call the *graph indifference*.

Let us now consider the set of all allowed distinct Leibnizian graphs that have precisely N vertices. They are, of course, finite in number. For each distinct graph we can calculate the graph indifference. Since this is a positive number, there must be a certain number of graphs, or perhaps just one, for which this graph indifference is smaller than for all the other allowed graphs. For simplicity, let us assume just one graph has the lowest indifference. I shall call this the *maximal-variety graph*, or *best graph*. It is the graph in which, in a well-defined sense, the vertices are more varied, more readily distinguished, than in any other allowed graph of the given number N of vertices. The requirement of minimal indifference selects – one could even say calls into being – a graph of vertices that 'strive' to be as individualistic as possible.

If, following Leibnizian epistemology, existence is identified as the possession of attributes that distinguish – the possession of positive variety – then a
world in which such distinction is done efficiently is surely to be preferred to
one in which the job is done less well – indifferently, indeed. Leibniz, of course,
always asserted that we live in the best of all possible worlds, but he was never
very explicit about how 'best' should be defined. Perhaps the closest he came
to expressing a criterion that admits precise mathematical formulation was in
Sec. 58 of the *Monadology*, in which, speaking of the "mutual accommodation
of all created things to each other," he says that this "is the means of obtaining the greatest variety possible, but with the greatest possible order; that is
to say, this is the means of attaining as much perfection as possible."

This passage was the stimulus to the development of the notion of a
maximal-variety graph. It should be emphasized that variety can be defined
in many different ways, and there is nothing sacrosanct about the particular
choice made here. (It was actually proposed by David Deutsch (1989), and
is different from the original proposals made by Smolin (1988) and myself
(1989); it seems to me to have the virtue of being extremely close to Leibniz's idea that every contingently existing thing is defined by the listing of its
attributes.)

3. Maximal Variety as Generator of Structure

There are many things I could say about this model, but my prime concern
here is with the generation of structure as a first principle. I shall therefore
refer the reader to the papers by Smolin (1988) and myself [Barbour, (1989),
(1991), Barbour and Smolin (1993)], in which we discuss possible applications
of the notion of maximal-variety graphs to theories of the origin of space and
time, alternative forms of quantum theory, and also the more detailed reasons
why we believe the notion of a maximal-variety graph does realize many of
Leibniz's ideas.

Turning to the question of structure, I think it is obvious that a maximal-
variety graph does indeed generate structure out of nothing – and, moreover,
as its first principle. As presented here, the maximal-variety graph (or graphs,
if more graphs than one have the same lowest indifference) is uniquely determined by just two conditions: 1) the law used to define the indifference (the
"reciprocal" of variety) of each of vertex and, hence, the indifference of the
complete graph; 2) the number N of vertices that the graph shall possess.
However, it would not be difficult to formulate a maximal-variety optimization problem in which N is to be determined as well. For this it would merely
be necessary to define the specific indifference of the graph (i.e., the indifference per vertex) and then require that the realized graph shall have the
lowest possible specific indifference. Since we have the freedom to modify the
definition of indifference (or variety) and also the manner in which the total indifference is normalized to find the specific indifference, it surely should

not be beyond the wit of mathematicians to devise laws that bring forth a maximal-variety graph for which its number N of vertices is also determined by the extremalisation principle.

With this modification of the generating law, one can say that the structure expressed by a maximal-variety graph represents pure law. By itself, the law creates the structure *ex nihilo*.

But what would such a structure look like? Unfortunately, Smolin and I have done very little work as yet on the 'pure' form of maximal-variety theory, i.e., in the fundamental graph-theoretical form. However, we have done some computer calculations on a less ambitious model that uses a residual pre-existing spatial framework to generate structure, and I would like to say something about that.

For simplicity, consider a closed one-dimensional world in which N sites can be occupied. One can think of a wheel with N slots on its rim into each of which one black or one white ball can be placed. Suppose this has been done; let W be the number of white balls, B the number of black: $W + B = N$. For each pair of distinct sites i, j we form the *relative indifference* I_{ij} as follows: We attempt to distinguish positions i and j by purely intrinsic denominations. For example, we might say that both neighbours of i have the opposite colour to i, whereas one of j's immediate neighbours has the same colour as j. Under such circumstances, i and j will be 1-step distinguishable and we define I_{ij} to be 1. However, it might be necessary to extend the comparison of the neighbourhoods of i and j to n steps before the two positions can be distinguished; then $I_{ij} = n$. In making the comparison of neighbourhoods, we are not allowed to distinguish by means of colour, i.e., by saying a neighbour is black, or by using left or right (by referring, say, to the neighbour on the left). By virtue of these two restrictions, the relative indifference will be defined in a left-right invariant and colour invariant manner. The formal algorithm for determining I_{ij} is this. Imagine j and all its environment shifted round to bring site j on top of site i. If j has the opposite colour to i, reverse it and its complete environment. Then see how many steps from the common position of i and j it is necessary to go before the environments become different; note the number n_1. Now reflect the environment of j about j and repeat the exercise, obtaining a new number n_2. The larger of these two numbers, or the common number if they are equal, is then equal to I_{ij}. The complete set of relative indifferences I_{ij} forms a symmetric matrix (whose diagonal elements we set equal to zero by definition). We can now form a total indifference in two different ways: either by finding the maximum value of I_{ij} in each row $i = 1, 2, \ldots, N$ and adding these N largest indifferences to give a total indifference (this is exactly analogous to the procedure I described in the pure graph model) or by simply adding all the $\frac{N(N-1)}{2}$ relative indifferences, which gives another number that may equally well be called the total indifference of the configuration. (Since $I_{ij} = I_{ji}$, one can make the summation subject to the restriction $i < j$ to avoid double counting.) The calculations I want to describe were actually made using this second definition of indifference.

With a good modern personal computer, it is quite easy to calculate all possible configurations having a given number of slots up to about 30 and thereby establish which have the lowest indifference for each given number of slots. I, in fact, did exhaustive calculations up to $N = 27$ with a rather pedestrian programme. The programme runs for about two days on my Macintosh SE/30. Increasing N by one doubles the number of configurations that must be examined and increases the time required to calculate all N-slot configurations by a factor somewhat greater than 2 (because the configurations are larger as well as being twice as many in number). My estimate is that, with a really smart programme, one might be able to calculate all configurations up to about 40 slots before a power cut crashed the run. However, the exponential rise of the configuration number with N will make it impossible to take exhaustive calculations much beyond that. On the other hand, by iterative trial-and-error methods it should be possible to find configurations with N of the order of several hundred and a very high variety (low indifference). However, as yet I have not made any such calculations.

Let me now describe the results that I did obtain. One can readily check that for $N < 7$ there are no Leibnizian configurations at all in the model I have described. They all contain at least two sites that are absolutely indistinguishable in accordance with the rules formulated above. For N = 7, there is just one Leibnizian configuration, which takes the form

$$\text{xxoxooo}, \tag{1}$$

which must be imagined bent into a ring with the first x (representing black) next to the last o (representing white; I choose to call the colour with fewer sites black, but that is purely nominal in view of the black-white symmetry). In accordance with the rules, we are to regard the configurations ooxoxxx and xoxxooo, for example, as identical to (1).

The configuration (1) can be regarded as the first solution that exists to our optimization programme for structure generation. It turns out that it anticipates all the minimum-indifference configurations that exist up to $N = 27$. As N is increased, the number of distinct configurations for given N increases quite rapidly, and for $N = 27$ is of the order of 1.2 million. I have found that, with few exceptions, the minimal-indifference configuration for a given N is not unique, though the number of such configurations is not large; for example, for $N = 14$ there are nine, for $N = 15$ there are three, for $N = 22$ there are four, and for $N = 27$ there are over 20.

The table which follows gives all minimal-indifference configurations for $N = 21$ to $N = 25$, which happen to be conveniently few in number:

N	Configuration
21	xxxooxoxoxoooxooooooo
22	xxoxoxoxxxxoooxoooooooo
	xxoxooxooxoxoxoooooooo
	xxxoxxxxooxoxoxoooooooo
	xxoxoxoxxxoxxoxoooooooo
23	xxxooxooxoxoxoooxoooooooo
24	xxxooxooxoxoxoooxooooooooo
25	xxoxxoxoxoxxxooxooooooooo
	xxxoxxxxooxooxoxoxoooooooo

Without any exceptions, the minimal-indifference configurations up to $N = 27$ all possess certain very characteristic features. First, about one third of the configuration is occupied by a uniform run of sites of all the same type. As represented above, these are the strings of zeros on the right of all configurations. I shall call this uniform run *the space*. In all cases, the space is bounded at one end by a single site of the opposite type followed by another site of the same type as the space. At the other end of the space, there are always two or three sites of the opposite type. After that the two types of site seem to alternate in a manner that is very hard to predict; without doing the exact calculation it seems to be impossible to say what will be found in the part of the configuration that is not the space and which I shall call *the body*. However, the body is always asymmetric, having two different ends. That much seems to be predictable.

As a very simple model of the world and its evolution in time, let us suppose that the passage of time corresponds to the *creation of possibilites*, represented by an increase in the number N of slots that can be filled in such a model. Then the first instant of time corresponds to $N = 1$, the second to $N = 2$, and so forth. At each instant of time, the world is required to fall into the minimal-indifference configuration for that slot number – it is condemned to be creative forever and always to seek the maximal-variety configuration. Then the above table of configurations shows us the appearance of such a world in the time interval 21–25. We see a space and a body that evolve and grow in what seems to be a deterministic manner as far as the gross structure is concerned but a probabilistic manner as regards the fine detail of the interior of the body.

But this is exactly the kind of behaviour that we observe in the world around us, in which the gross structure seems to evolve in accordance with deterministic laws of classical physics, while the microscopic structure seems to obey probabilistic quantum laws. This sort of result encourages me to think some form of theory based on the ideas of maximal variety could provide a realistic model of the world, which would therefore be described by a basic law completely different from either quantum or classical physics but one that nevertheless is capable of reproducing the results of both of these theories under certain well-defined conditions.

4. Leibnizian Aspects of Maximal Variety

Although the idea of maximal variety was developed mainly with the (long-term) aim of creating new physical theories, it does seem to me interesting in its own right as a mathematical model of the *Monadology*. I have written at more length on this subject elsewhere (Barbour 1993), but would here like to mention two intriguing aspects of the model.

The first is the observation that if the ideas behind this model are on the right track, physics will become much more like biology than it is at present. For all the entities in the model are created in a kind of ecological balance between competing individuals. Each is trying to be as individualistic as possible, but in a curious way this very selfish behaviour is necessary if anything is to exist (become differentiated) at all. By making ourselves differentiated, we have to make other beings differentiated at the same time. Surely Leibniz would like this example of how seemingly bad (selfish) behaviour is needed to make even the best of all possible worlds. *Vivat Pangloss!*

I shall spend rather more time on the second Leibnizian aspect of the model: it is the possibility of constructing a completely idealistic (in the Berkeleian sense) model of the world in precise mathematical terms without lapsing into solipsism.

For this purpose, I shall assume that a maximal-variety graph encodes the state of the world at one given instant (there are various ways I can think of in which time could be brought into the scheme – one was just outlined – but as yet none seem to me satisfactory). We take each vertex to represent a sentient being, a monad. Then for a given vertex-monad the remaining lines and vertices stand for two things simultaneously: 1) the relations that monad has to all the other monads at that instant and 2) the perceptions experienced by that monad at that instant. According to Leibniz, the two are really the same thing. Under 1), I would understand things in rather abstract mathematical terms. Thus, a line joining two vertices i and j will mean that i and j have a direct, or first-order, relation to each other; one could also say that they experience each other's existence directly. If j is joined to k, but k and i are not joined, then we can say that k and i have a second-order relation. But one could also give the bare mathematical structure a more concrete meaning (2) by saying that a line from a given vertex i to a vertex j that has no other connections means that vertex i has some definite experience, say being aware of the colour white. If j is connected to one other vertex, that could mean i experiences the colour blue. For each definite type of vertex to which connection is made, there could correspond a definite sensation (the precise sensation experienced could depend on quite remote connections within the graph). Given such a lexicon – the translation from the bare elements of the graph to actual experiences – we could read off the experiences of each of the monads within the graph.

Note that although the fact of being joined by a line is a reciprocal relation, so that if i has an awareness of j we must also assume that j has an awareness of i, what is actually experienced will, except in rare cases, be different, since the other connections from i and j are not the same, and, by hypothesis, it is these other connections that determine how each vertex is experienced.

It is interesting to consider in the framework of this scheme Leibniz's famous remark that monads have no windows through which attributes might enter or leave a monad (*Monadology*, Sect. 7). This is trivially true in the present model; for in the bare graph-theoretical terms a given vertex is nothing more than the listing of its connections to other vertices. These connections are its attributes and it has no others. The attributes of a vertex could only be changed by considering a *different* (but very similar) graph in which some of the connections have been changed. But, strictly, that is then a quite different world and consists of different monads. (Since we have assumed that the actually realized world is the one that exhibits the greatest possible variety, this modified world will possess less variety and hence fail to make it into existence – it will be one of Leibniz's *possible worlds*.) Thus, each monad's attributes in the actually realized world are given once and for all. Each monad is, in fact, simply *the world as seen from its particular point of view*.

Leibniz was wrong to think that in a monadological world the mutual consistency of all the different monads – what he called the pre-established harmony – is a great miracle of God. In fact, it is a trivial consequence of the model. (The miracle is that *anything* has been created and is experienced.) In a graph, each vertex is willy-nilly a view of the whole from a given point of view. The graph defines all the views and enforces their mutual consistency. Moreover, it is inherent within the model that the whole consists of and defines its parts and that each part is simultaneously the whole.

In logical terms it may be true that the monads have no windows. But in another sense they are literally riddled with windows. What I as a particular vertex experience is of necessity a function of what the other vertices experience. The experiences are not the same but they are still related. Once I have achieved full self-awareness and understand in graph-theoretical terms why I have particular experiences – the colour yellow, say, because I have a first-order relation to a three-line vertex – I will simultaneously know something about the experiences of the vertex which is responsible for my experiencing yellow. Increasing self-awareness (gained from awareness of the connections between what we observe) of the connections within the graph of which we are a part, which simultaneously gives us the lexicon for translating graph-theoretical concepts into direct experiences, simultaneously enables us to deduce some at least of the experiences of the vertices to which we are more directly related. We can *peep* into the experiences of our fellow monads. Because of the way in which experiences are generated, we are all continually sharing experiences, though there is never identity of experience.

In fact, the entire world is resolved into *pure shared experience*. This is an appropriate place to stop. If my conjecture is correct, this idea must resonate within you.

References

Barbour, Julian B. (1989), Maximal variety as a new fundamental principle of dynamics. Foundations of Physics, **19**, pp. 1051–73.

Barbour, Julian B. (1993), Mathematical modelling of the Monadology. Submitted for publication.

Barbour, Julian B., and Smolin, Lee, Extremal variety as the foundation of a cosmological theory. To be published in Classical and Quantum Gravity.

Deutsch, David (1989), Private communication.

Galilei, Galileo (1632), Dialogo Sopra i due Massimi Sistemi del Mondo Tolemaico e Copernicano, Florence.

Kepler, Johannes (1596), Mysterium Cosmographicum, Frankfurt.

Prigogine, Ilya (1980), From Being to Becoming. San Francisco: W. H. Freeman & Co.

Russell, Bertrand (1945), A History of Western Philosophy. New York: Simon & Schuster.

Smolin, Lee (1991), Space and time in the quantum universe. In: *Conceptual Problems in Quantum Gravity. Proceedings of the 1988 Osgood Hill Conference*, Ashtekhar, A., and Stachel, J. (eds.), Boston: Birkhäuser.

Galaxy Creation in a Non-Big-Bang Universe

Halton Arp

Max-Planck-Institut für Astrophysik D-85748 Garching, Germany

The Big Bang

The most widely accepted model of the universe today is the Big Bang. The reason most people readily accept this model is that they believe that the recession of the galaxies from us requires the universe to have originated in a single cosmic event. It was this inferred recession of galaxies after all which suggested more than 60 years ago that our universe originated from an earlier, compressed state. But the observations require not just the recession of all galaxies away from our own, they require the recession of all galaxies away from all other galaxies – the space in which the galaxies are imbedded is supposed to be expanding. Stated in this fashion an alternative possibility becomes immediately obvious: The galaxies could be continually created at many points within an expanding space. For example Bondi and Gold[1] and Hoyle[2] in 1948 postulated continuous creation of hydrogen atoms between the galaxies which kept the density constant (steady state) in an expanding universe. Therefore the creation of all galaxies at a single instant of time, as required by the Big Bang, is not necessarily required by an expanding universe.

A spirited debate sprang up between adherents of the Big Bang and steady state theories. The exact form of the redshift-apparent magnitude (Hubble) diagram was subsequently supposed to have been decided in favor of the Big Bang. But then it was realized that galaxies must evolve (change their luminosity and color) as they aged over the long interval since their creation in the Big Bang. That invalidated the arguments about deviation from linearity of the Hubble diagram (The arguments involving the so-called deceleration parameter q_0). The argument was then circularized to say that the (assumed) evolution of the mean characterstics as we looked back in time ruled out the steady state and ruled in the Big Bang. But eventually it was shown that,

---

[1] (Bondi, Gold 1948)
[2] (Hoyle 1948)

for example, the number counts of radio sources were not necessarily greater at earlier times. It depended on what direction in the sky one looked and in addition, certain classes of objects like radio galaxies showed constant density, Euclidean geometry behavior as a function of decreasing apparent magnitude (Das Gupta, Narlikar, Burbidge 1988).

In reviewing the arguments after some years I remember Fred Hoyle remarking: "Of all the supposed arguments disproving the steady state the only one remaining as serious evidence is the cosmic background radiation – interpreted as relict radiation from the Big Bang at $T \approx 2.7\,°K$."

Now, however, the measurements of the CBR have been giving increasingly smooth results. There is no trace of the fluctuations from point to point in the sky which should be evident if galaxies formed in the early phase of a Big Bang universe. The specialists say that the CBR is worryingly smooth – and if it gets any smoother the Big Bang will be in real trouble. A few years go by and they then report it has indeed gotten smoother and that if it gets any smoother the Big Bang will be in real trouble, etc., etc. I think it is the height of irony that the last remaining proof of the Big Bang now becomes evidence against the Big Bang.

Another dark cloud has long been building on the horizon of the Big Bang. Over the last 23 years evidence, and some debate, has been growing as to whether all cosmic redshifts are caused by Doppler recession velocities. Some opponents of the Big Bang have been encouraged by the reasoning that if some redshifts were non-velocity that all redshifts might arise from another cause. In other words the universe might be static. My own opinion, however, is that even if someday many cosmic objects were shown to have a majority of their redshift intrinsic, that it would still be difficult to disprove the hypothesis that they had at least some small, systematic expansion – that there was not some underlying dilation of space. So with the present state of the evidence the expansion or non-expansion of the universe is irrelevant to the key cosmogonic question of the Big Bang, namely: Did the universe originate at a single point in space time?

But this operational definition of the Big Bang turns out that to be precisely where we can observationally test the hypothesis. The crucial question is simply: Are all galaxies old? The most profound aspect of the Big Bang is that it represents one single act of creation of everything. All galaxies are supposed to have formed $\sim 2 \times 10^{10}$ years ago. But it is readily apparent to even non-specialists that we see many galaxies filled with young, blue stars. The obvious conclusion that these galaxies were recently created was dismissed a long time ago with the argument that all galaxies formed about the same epoch in the early universe but that some have been forming stars more slowly than others. In principle this argument says you might have a galaxy with most of its stars obviously $\sim 10^8$ years old but containing only one star $\sim 2 \times 10^{10}$ years old. Thus it is difficult to observationally prove any given galaxy is not $\sim 2 \times 10^{10}$ years old. This is the key argument underpinning Big Bang genesis which has so far avoided critical evaluation.

I would like to point to evidence which refutes this key requirement of the Big Bang. It is simply that if galaxies all formed $\sim 2 \times 10^{10}$ years ago, and that star formation proceeded more slowly in some, that today we should see galaxies in all stages of apparent youth, including many galaxy sized masses of hydrogen gas that have not yet started forming stars. We do not.

Galaxy size masses of hydrogen are easily detectable out to considerable distances in the universe, particularly in recent decades with sensitive radio telescopes. An important review of this subject was given by Morton Roberts (Roberts 1987). He reported "Isolated hydrogen clouds have been sought for in many experiments. None have been found ...". A search for hydrogen clouds reported in 1989 that in the redshift range $z = 3\,000$ to $10\,000\,\mathrm{km\,s^{-1}}$ no definite or even probable clouds were found (Brosch 1989). In another search at redshift $z = 4$, where galaxy formation should be very active on the Big Bang hypothesis, galaxy sized clouds of hydrogen were not found.

In essence the argument is that if galaxy sized hydrogen clouds were all born $\sim 2 \times 10^{10}$ years ago that they would have had to, one by one, flare up in episodes of star formation between that epoch and the present time. But this peculiar behavior would have had to suddenly stop at just our particular time in the universe at which we are observing. (Since there are no more proto-galaxy sized clouds.) This is so unlikely a circumstance as to, in my opinion, constitute a disproof of the Big Bang.

Recently much publicity has been given to the discovery of a hydrogen cloud in the Virgo Supercluster galaxies (Giovanelli, Haynes 1989). It has been suggested (rather nervously in view of the Big Bang) that this is a "young" galaxy which has not yet started to form stars. But, as we have just argued, it is actually the general absence of such delayed formation galaxies which require the many young galaxies we observe to be indeed newly created. The key property of this newly discovered hydrogen cloud is that it is only $\frac{1}{10}$ the mass of our own Milky Way Galaxy. Our own galaxy is only a fraction as large as the big galaxies which make up the universe. Therefore this newly discovered cloud is only a tiny companion, like those clouds ejected or torn off active galaxies or galaxies in groups (Arp 1985). The importance of this observation is that after all the searches which failed to uncover galaxy sized masses delayed in formation from the Big Bang, that the candidate which finally turns up is only about $\frac{1}{100}$ the required mass. It is rather clear that the much larger clouds of the size that are needed to save the hypothesis of the Big Bang would easily been discovered if they were there.

For direct evidence let us examine actual young galaxies. Do they look as if they were ancient gas clouds suddenly now condensing stars throughout their volume? On the contrary – they are characteristically centered on a small, energetic core out of which most of the material in the galaxy appears to flow. One example is 3C 120, a compact galaxy originally classified as a quasar, which has optical jets, young stars and extensive radio jets emanating from its condensed, active nucleus (Baldwin et al. 1980, Arp 1987a). A radio map superposed on a photograph of this object is shown in Fig. 1. It is a good illustration of the general tendency of galaxies to eject radio material out into space in roughly opposite directions. Evidence of luminous matter ejected in these directions is also observed (note peculiar galaxy in northern radio jet – it is about $5\,000\ \mathrm{km\,s^{-1}}$ higher redshift).

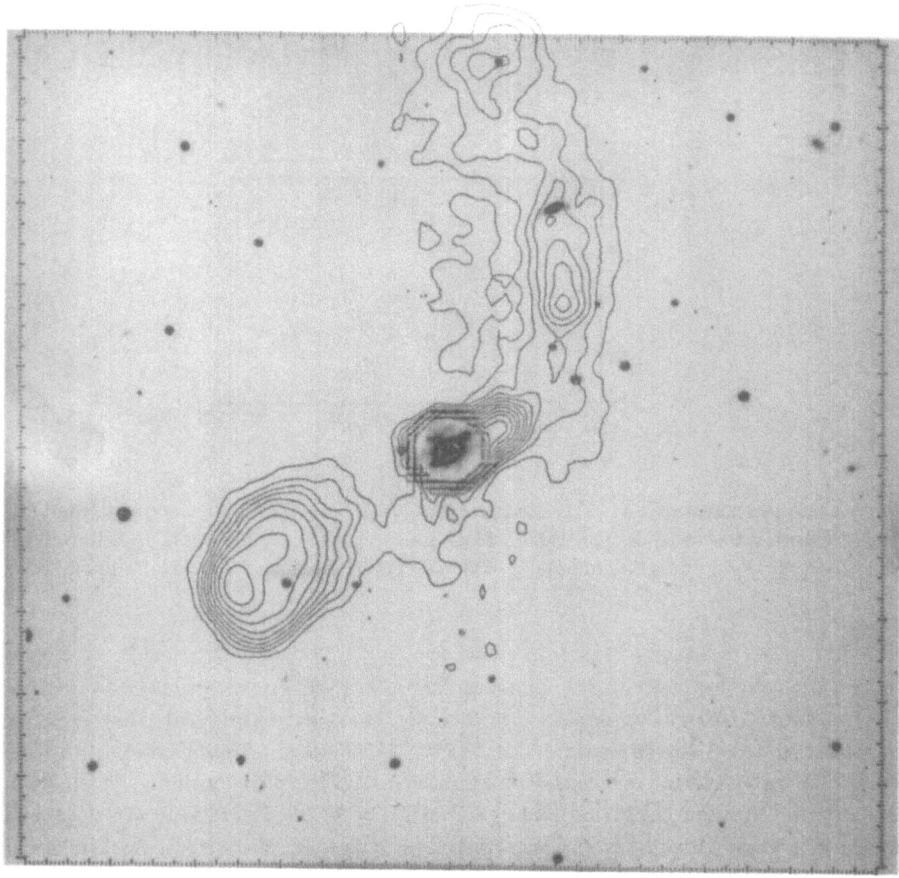

Fig. 1. The active, compact galaxy 3C 120 was originally classified optically as a point source quasar. Typically for such objects radio material is streaming out of the nucleus. (Shown by isophotos). Good seeing, large reflector photographs show optical jets in the interior pointing out in direction of ejection. A peculiar companion galaxy of $5\,000\ \mathrm{km\,s^{-1}}$ higher redshift is visible in upper right radio lobe. (Photograph by H. Arp)

In order to avoid the conclusion that significant matter emerges from these active nuclei, conventionally astronomers argue that only energy is ejected, en-

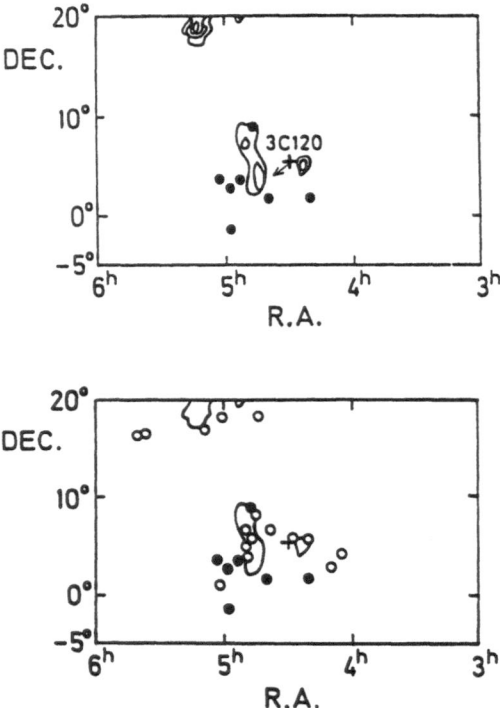

Fig. 2. Centered around 3C 120 is a cluster of quasars ($z \sim 2$) as shown by filled circles. Open circles show concentration of galaxies with $z \sim 5\,000\,\mathrm{km\,s^{-1}}$ and open areas show neutral hydrogen clouds ($z \sim 0\,\mathrm{km\,s^{-1}}$). 3C 120 itself has $z \sim 10\,000\,\mathrm{km\,s^{-1}}$.

ergy which dissipates with few consequences. They argue old material falls into the nucleus to cause explosions, or nearby galaxies collide with this central engine and "feed the monster". That is why it is so important to examine the evidence for material ejection from galaxies. Aside from evidence of ejection of compact bodies called quasars (Arp 1987b) there is evidence of material ejection from galaxies which are conventionally accepted to be in the most active stage of star formation. A non-controversial example is the "starbursting" galaxy NGC 253 (Ulrich 1972, 1978, Arp, Sulentic, preprint). It has been shown that outflow of material from the center is causing rapid star formation. No colliding or infalling companions are responsible for this spontaneous activity.

If we are forced, by the earlier argument that all galaxies cannot be old, to accept that these active blue galaxies are more recently created – then we are forced by the evidence to say they are created at some small central point and flow outward over time to form the new galaxy. We have now arrived at the empirical alternative to the Big Bang, namely: continuous creation.

Theoretically this situation is not forbidden, in fact it has been strongly suggested for a long time. More than 40 years ago P.A.M. Dirac hypothesized additive or multiplicative creation of matter in the universe. (Multiplicative would be the creation of matter enhanced by the presence of other matter – such as in active galaxy nuclei). Fred Hoyle (Hoyle 1960) formulated in 1960 the C field ("C" for creation) into the general relativistic equations which the universe must satisfy. Much later Alan Guth (Guth 1981) described fluctuations in the "material vacuum" which created mass. He and Andrei Linde (Linde 1987) followed the implications of creating "baby universes". Gunzig, Géhéneau and Prigogine (Gunzig et al. 1987) reaffirmed the validity of introducing a mass creation term in the Einstein geometry-energy tensor. It would seem that it has taken a long time for the obvious idea to occur to people: If a Big Bang happened once, why not again? And again? The Big Bang idea, which is not a scientific theory because its origin is excluded from observational verification, is now capable of scientific investigation because we have the possibility of observing mini- bangs in different stages of development in different places in space.

Conventional Evolution

When extragalactic objects were observed at higher redshift (assumed to be greater distance) their properties turned out to be increasingly different. For example for redshift $z > 1$ the Hubble relation turned sharply upward from linearity (Spinrad, Djorgovski 1986). This had to be attributed to evolution (galaxies, roughly speaking, were brighter in the past). Of course that interpretation conflicted with the continuous star formation scenario which was used to explain currently young appearing galaxies. Also, as Joe Wampler pointed out (Wampler 1986), quasars which ostensibly had continua arising from synchrotron plasma processes showed the same supposed evolution as galaxies with continua arising from stellar aggregates – a quite impossible circumstance. But these contradictions were ignored and the general concept of evolution was used as an adjustable parameter to avoid contradiction with the observations. The general difficulty with this picture was that it required a universe which changed its characteristics drastically over relatively short periods of time. The universe which was treated in the space-time of general relativity as absolutely homogeneous over large spatial dimensions was then utterly inhomogeneous over large time dimensions.

As we remarked, the adjustable parameter of "evolution" which was used to reconcile the observations with the theory was then used to "disprove" the steady state theory. But recall a little emphasized circumstance; the steady state theory was simply a special case of continuous creation theory. Continuous creation might be uneven and hence homogeneity on any arbitrary scale might or might not be expected. Therefore an assumption about evolution had been used to rule out an alternative theory to the Big Bang – an

alternative theory which was not even representative of the general class of
alternative theories. As we shall see, it is an intriguing question as to what
scale size, if any a continuous creation universe would become repetitive, i.e.,
homogeneous.

Empirical Evolution

The empirical picture of evolution is derived simply by studying galaxies physically associated together. It is clear that the largest galaxy in such groups
typically contains the oldest stars and the smaller, higher surface brightness
and generally more active galaxies contain the youngest stars. This evidence,
going back almost 30 years, is extensive (Arp 1987b). The difficulty is that
the redshifts of these group members increases steadily until the smallest, (not
at their z distance) active, very high redshift quasars are encountered. The
association of different redshifts together has prevented general acceptance of
the groupings because many influential extragalactic astronomers insist that
redshifts can only be caused by Doppler velocities of recession and that high
redshifts must place the more active members of the group at varying distances
in the background where their association is only apparent. The astronomers
who do accept these physical associations point to not only overwhelming statistical evidence but also the observed physical interaction of high and low
redshift objects – and particularly certain peculiar, compact high redshift objects attached to ejecting parent bodies by luminous filaments.

On this latter interpretation the key point is that the young galaxies
are small and compact and emerge from small active nuclei of older galaxies. The observed sequence of forms, correlated with the decreasing amount
of intrinsic redshift, empirically suggest an evolution from a small, compact
quasar, through active compact galaxy, through spirals of increasingly early
type ($Sc \rightarrow Sa$) to the largest galaxies like the E's and S0's which are pre-
dominantly filled with old stars.

If this evolutionary picture is valid the protogalaxy which emerges from
the nucleus of an older galaxy must grow in size with time (required because
the very active nucleus of an ejecting galaxy is small relative to a partially
evolved companion galaxy). Galaxies dwindling in mass as they become older
is opposite to what we observe so we are forced to the conclusion that the
material is created in the interior of the older galaxy and expelled to grow
into a new galaxy which ages and grows larger and presumably undergoes
secondary ejection-creation at some stage.

But this is just the continuous creation picture of galaxies which we arrived
at in the beginning of this review from very general considerations of galaxies
with young stellar populations, the lack of proto-galaxy hydrogen clouds and
the contradiction to expectations of what should be observed on a Big Bang
hypothesis. It is impressive to me that the empirical relationships induced
from the detailed observations of individual objects lead back to the same

alternative to the Big Bang as required by the generally observed abundance of young galaxies.

Creation

At first glance it would seem difficult to prove or disprove the proposition that new matter was created in the interiors of active galaxies. We cannot observe over a significant span of time what happens in or around nucleus. We can only infer the stages from our empirical, evolutionary sequence. But it turns out fortuitously there are two properties of the young galaxies which are unexpected and baffling but nevertheless, in a very general way, argue for a true creation process.

Both of these properties have to do with the redshifts of the supposed younger galaxies. These redshifts are systematically positive and increase as the objects become more compact and quasar-like. One must accept that these redshifts are non-velocity, intrinsic to the material of the body and not caused by gravitational red-shifting or tired light from particle interaction in the vicinity of the red-shifted body. Many astronomers do not concur with the alternate "young material" explanation. I can only say that I think the nature of the observational evidence requires it (Arp 1987b).

Given the conclusion that the redshifts are a property of the material making up the galaxy, however, I would argue that they could only be produced by the creation of zero mass matter. As the matter ages and each proton and electron acquires increasing mass, the mass of the electron in the Rydberg of the radiating atoms causes the redshift to decrease. The objects start out compact, young and with high intrinsic and shift. They grow and expand through the sequence of galaxy forms we have described and as they age they are characterized by lower and lower intrinsic redshifts. The initially very high intrinsic red shift could only come about if the matter materialized from an initially zero mass state. This defines what would be a sensible meaning to the operation "creation" – viz: *Matter which was in a non-localized state in the Universe materialized in a given locality by growing from mass zero to an observable value.* If the observations are correct in requiring evolving, intrinsic redshifts for quasars and galaxies then, I would suggest, only matter creation could explain it.

Quantization of Redshifts

There is one other startling property of extragalactic redshifts which I would suggest requires continuous creation. This property is so unexpected that most astronomers stubbornly resist accepting it even though the evidence has been building up now for about 20 years. It is the periodicity, or quantization, of extragalactic redshifts.

Figure 3 here shows that *all* extragalactic redshifts, from the largest redshifts associated with quasars to the redshifts of the most normal galaxies in small groups are quantized. The preferred values of quasar redshifts were first noticed by Geoffrey and Margaret Burbidge in 1967[3]. The formula which fit their periodicity was discovered in 1971 by K.G. Karlsson[4]. The small periodicity of $72\,\mathrm{km\,s^{-1}}$ was discovered in 1967 by William Tifft in binary galaxies.

Now for many years astronomers have been measuring redshifts of galaxies in various regions of the sky. It has long been rumored that they were finding puzzling groupings in these redshifts. Finally they took courage in joining together and announced that the combination of four different surveys showed inescapable periodicity in the red shifts (Broadhurst et al. 1990). They insisted on expressing this as a spacing distance of $128\,h^{-1}\,\mathrm{Mpc}$. But the actual observed data are redshifts. Only the graphical data on redshifts has so far been published but the largest peaks are estimated by me to be at $z \sim .3, .24, .18, .12$ and $.06$ (the latter for the folded North-South data in the bright surveys). These peaks are illustrated here in Fig. 3. If I am correct in the assignment of these particular preferred redshifts to galaxies then the result must be another key to the origin of galaxies. As for the significance of the periodicity it means that the two lowest redshift peaks observed in the quasars at 0.30 and 0.06 *are also observed in galaxies in general.* This is not surprising because quasars, particularly low redshift ones, are undeniably a kind of galaxy. But it reinforces enormously the reality of these discrete extragalactic redshift peaks because they are found independently in very different sets of data!

The smallest periodicity of $72\,\mathrm{km\,s^{-1}}$ is well established (but clearly not well accepted) in groups and pairs of galaxies and can be traced as high as $288\,\mathrm{km\,s^{-1}}$, possibly as high as $360\,\mathrm{km\,s^{-1}}$. In the Virgo cluster and Local Group (Napier, Guthrie 1988) the $72\,\mathrm{km\,s^{-1}}$ periodicity can be traced to a thousand or more $\mathrm{km\,s^{-1}}$. This appears to mean that as extragalactic redshifts become smaller, the periodicity exhibited by the redshifts becomes finer and finer. But in the overall view of redshifts the *whole of the observed range contains preferred values* and submultiples thereof. In other words the red shifts are quantized.

---

[3] (Burbidge, Burbidge 1967)
[4] (Karlsson 1971, 1977)

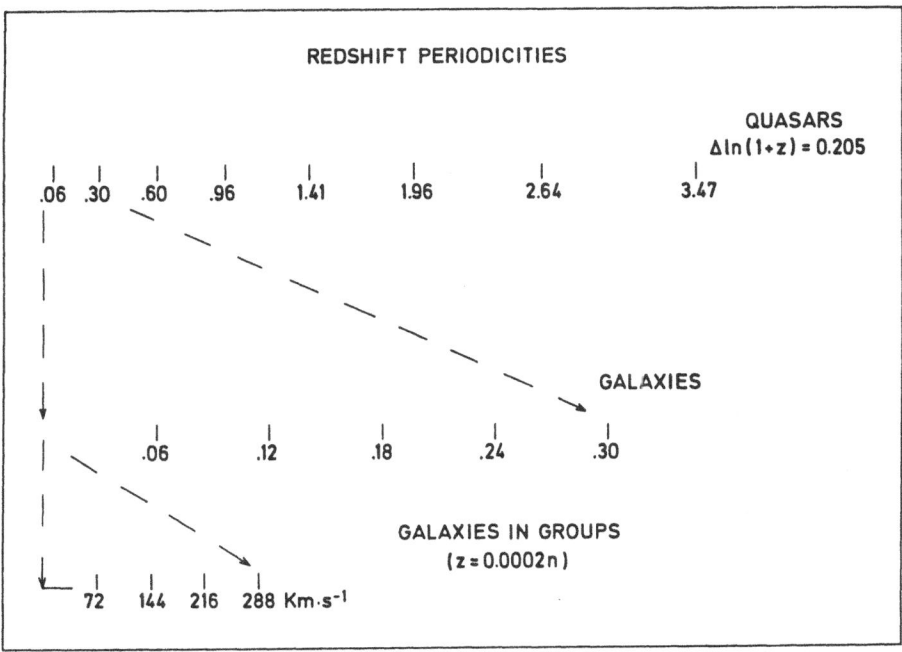

Fig. 3. Summary of known redshift periodicities showing entire red shift domain is quantized, and in increasingly fine steps as redshifts decrease. Redshift peaks are as discovered by various investigators from independant data as discussed in the text.

In general I would argue that this requires an origin of galaxies from a zero mass/energy state (Arp 1989). Where else could matter receive discretized, intrinsic properties? We could only venture some speculative questions about such a mechanism.

How do the fundamental particles making up the matter of a galaxy acquire increasing mass as time goes on? My suggestion is that as they exchange gravitons with an increasing volume of their newly visible universe. The universe they can communicate with has a radius of the velocity of light times their age since they were created. Therefore their intrinsic redshift decreases as they age. But because the matter materializes in a zero mass/energy state it must emerge from a quantum mechanical domain. Therefore some fundamental property must be quantized, for example, Δt, the time between creation episodes. (Which could be viewed, alternatively, as discrete mass differences between succeeding creations). Would the redshifts decay uniformly, keeping their initial spacings? Not unless the increasing volume they communicated with was homogeneous. If there were lumps or shells in that volume the redshifts would decay in discrete steps. One predicted consequence, however, would be that as the volume embraced became larger the redshifts would become smaller and the steps in redshift would become smaller as observed.

The observations of quasar redshifts show there are small phase shifts from region to region of the sky (Arp et al. 1990). Would further study of these give information on the size and nature of our local creation event?

These suggested details surely cannot have much chance of being correct. They do serve, however, to give a counterexample to the statement that "we cannot accept the observations because it is impossible to explain them". The suggestions made in the present paper are meant to illustrate that there are possible kinds of explanations within the observationally required framework of continuously created, new galaxies.

Note Added in Proof

The observational conclusions in the present paper received thoretical support in a recent paper by J. Narlikar and H. Arp (1993) in wich it was shown that a *general* solution of the equations of general relativity leads to a static (non-expanding) universe. The *special* solution which has led in the past to the expanding space of the Friedmann, Big Bang model rests on the assumption of particle masses being constant in time in the universe.

The new, more general solution requires (intrinsic) redshift to be a function of the age of its matter and, therefore, more distant or younger galaxies to be increasingly redshifted. Exactly the ovserved value of the Hubble constant is predicted (50 $\mathrm{kms^{-1}Mpc^{-1}}$) for the non-expanding universe. The creation points of new matter (mass $= 0$) are precisely the connection between quantum and classical mechanics and replace the embarassing space-time singularities in the conventional, mass $=$ constant solution of cosmology.

As for the cosmic background radiation, its interpretation is much simpler on this new continuous creation theory than in the Big Bang theory. For the expanding universe, the 2.7 K radiation can only come from a very thin, very distant shell where radiation decouples from matter. (Intermediate distance shells would redshift the radiation because of varying expansion velocity). But such a shell should show the irregularities due to primordial galaxies. The extreme smoothness which is actually still the most striking aspect of the present observations (one part in about one hundred thousand) of the cosmic microwave background is, on the other hand, very simply explained by averaging along the line of sight from the observer to the visible limit in a non-expanding universe. In this latter case we are simply observing the temperature of the intergalactic medium.

References

Arp, H. (1985), Astron. J. *90*, 1012

Arp, H. (1987a): J. Astrophys. Astr. (India) *8*, 231

Arp, H. (1987b): *Quasars, Redshifts and Controversies*, publ. Interstellar Media

Arp, H. and Sulentic: J.W. Max Planck Astrophys. Preprint 388.

Arp, H. (1989): *Extragalactic Evidence for Quantum Causality*, Aperion, No. 5, 5

Arp. H., Bi, H.G., Chu, Y. and Zhu, X. (1990): Astron. Astrophys. in press.

Arp, H. and Burbidge, G.R. (1990): Ap. J. *353*, 21

Baldwin, J.A., Carswell, R.F., Wampler, E.J., Smith, E.E., Burbidge, E.M., Boksenberg,A. (1980): Astrophys. J. *236*, 388

Bondi, E. and Gold (1948): T. M.N.R.A.S. *108*, 253

Broadhurst, T.J., Ellis, R.S., Koo, D.C., and S Zalay, A.S. (1990): Nature *343*, 726.

Brosch, N. (1989): Astrophys. J. *344*, 597

Burbidge, G.R. and Burbidge, E.M. (1967): Ap. J. Lett. *148*, L 107

Das Gupta, P., Narlikar, J.V. and Burbidge, G. (1988): Astron. J. *95*, 5

Giovanelli, R. and Haynes, M. (1989): Science, *245*, 933

Gunzig, Géhéneau and Prigogine (1987): Nature *330*, 603

Guth, A. (1981): Phys. Rev. D. *23*, 347

Hoyle, F. (1948): M.N.R.A.S. *108*, 372

Hoyle, F. (1960): M.N.R.A.S. *120*, 256

Karlsson, K. (1971): Astron. Astrophys. *13*, 333

Karlsson, K. (1977): Astron. Astrophys, *58*, 237

Linde, A. (1987), Phys. Today *40*, 61

Napier, W.M. and Guthrie, B.N.G. (1988), in "New Ideas in Astronomy" p. 191 Cambridge Univ. Press

Narlikar, J. and Arp, H. (1993): Ap. J. *405*, 51

Roberts, M.S. (1987): *New Ideas in Astronomy*, 65, Cambridge Univ. Press

Spinrad, H. and Djorgovski, S. (1986): I.A.U. Symp. *24*, 129

Subrahmanyam, R. (1989): Ph. D. thesis, Indian Institute of Science, Bangalore

Ulrich, M.-H. (1972): Astrophys. J. *1978*, 113

Ulrich, M.-H. (1978): Astrophys. J. *219*, 424

Wampler, E.J. (1986): I.A.U. Symp. *24*, 147

What Kind of Science is Cosmology?

Hubert F. M. Goenner

Institut für Theoretische Physik, Universität Göttingen, Bunsenstr. 9,
D-37073 Göttingen, Germany

1. Introduction

Possibly, a philosopher of science is expected to ask this question – not a
theoretical physicist. Nevertheless, an inquiry into the nature of a discipline
covering, on the scale of atomic time, a span between 10^{-44} seconds (Planck
time) and 10^{100} years (the time after which a supermassive black hole of galaxy
mass will have been radiated away through the Hawking process) seems called
for also from the point of view of physics[1]. In cosmology we encounter a field
of research well established by all social criteria, a common endeavor of mathematics, theoretical physics, astronomy, astrophysics, nuclear and elementary
particle physics claiming to explain more than the cosmogonic myths of the
days of old. Has cosmology become a natural science, even a branch of the
exact sciences?

In following J. Ziman (Ziman 1968) we define natural science as an *empirical* science steered by public agreement among scientists. In this context, "empirical" means that conclusions are not merely drawn by rational
thinking but that they are tested by help of reproducible quantitative exper-
iments/observations. These measurements are then interpreted by consistent
(physical) theories and receive a preliminary validation to be reconsidered
in the light of new data. A characteristic feature of research in the natural
sciences is the collection of precise empirical data and their connection by
selfc-onsistent theories. In consequence, technical applications, possible prediction of novel relations among the empirical data (new effects) obtain as
well as models of explanation and understanding for the systems investigated.
It is essential that such explicatory models map, with a minimum of hypotheses, a larger piece of the network of relationships found in the external world

---

[1] The Planck time is $t_{Pl} = (\hbar G/c^5)^{1/2}$. In the literature, the evolution of the universe
up to times much larger than 10^{100} years is considered. Cf. Discuss et al. 1985 or
Barrow and Tipler 1987.

into percepts of our mind. It is particularly important that we are led, by such understanding, to new possibilities of qualitative or, better, quantitative experimentation/observation.

Does cosmology, which according to its general definition in dictionaries (Webster 1963, Petit Robert 1985) is the science (of the physical laws) of the universe, fit into such a frame? In trying to give an answer I will review, in brief, three phases of cosmological modeling: the standard model, the early cosmos, and quantum cosmology. Einstein's theory of gravitation is the exclusive theoretical background adopted here not just because the length of my contribution is limited but also because the implications for cosmology of general relativity are developed best.

In the following I shall use the words "cosmos" and "universe" as synonyms although they carry different rings; cosmos goes well with order and coherence, while universe implies uniqueness and entirety. It should be clear, however, that I consider cosmology neither as the playground for a Theory of Everything nor as the Holistic Grail of sensitive minds. Before going into details of cosmological modeling I will try to circumscribe cosmology as a field of research.

2.   The Content of Cosmology

2.1   The Universe: an Evasive Physical System

Sciences or branches of science usually are classified by the subject investigated or by the methods of investigation used. Thus, cosmology could be called "cosmophysics" in parallel with geophysics or solid state physics because its subject is the cosmos. "In cosmology we try to investigate *the world as a whole* and not to restrict our interest to closed subsystems (laboratory, Earth, solar system etc.)" (Sexl and Urbantke 1983). We encounter similar definitions in many textbooks. The world *as a whole*, though, is not readily accessible, empirically. Authors therefore provide qualifying attributes: the *visible* universe (Rowan-Robinson 1977), the *physical* universe (Hawking and Ellis 1973), the *astronomical* universe (Mc Vittie 1961), the *observable* universe. We all are aware of the fact that the domain of nature observable to us depends on the power of the available measuring instruments and thus changes (expands?) permanently. Cautious authors avoid the word universe in favor of expressions as "the metagalaxy" (Alfven 1967) or "structure on a large scale" (Li Zhi Fang and Ruffini 1983) or "distribution of matter on the largest scale" (Buchdahl 1981). In my course I use the following definition: "We understand the universe to be the largest presently observable gravitationally interacting system". Although it satisfies the needs of the practicing cosmologist, from the point of view of epistemology such a definition is hardly acceptable. I doubt that the bootstrap definitions given by Bondi (Bondi 1961) are more helpful: "The universe is

(a) the largest set of objects (events) to which physical laws can be applied consistently and successfully;

(b) the largest set of all physically significant objects (events)."

The average physicist seems to be an utilitarian thinker not worrying about the domain of application of his theories: in the wake of time he expects to learn more. It is taken for granted that the physical system "universe" exists. That this may be only in the sense of a mathematical limiting process or of an ontological construct ("the largest inextendible entity") is a problem left by physicists to natural philosophers. Progress of research seems not to be hampered by this attitude. Another example is given by the concept of elementary particle in the sense of the smallest indivisible entity. At first, it should have been the atom, then the nucleus and, presently, it is the quark – with no end of further subdivisions in sight. While I agree that an approximative approach to the universe as a physical system may be the only one allowed to physicists, there is the danger that an approximate concept is taken for the real one. In fact, particularly in quantum cosmology, the universe is treated as an entity resembling more a particle among other particles than the totality of gravitationally interacting masses on the largest scale (cf. also Sect. 2.2). In a way, cosmophysics, methodologically, is *opposite* to phenomenological thermodynamics. There, valid laws are formulated without the need to know the detailed microscopic structure of matter. In cosmophysics, we are dealing with the detailed knowledge of structured parts of a system unknown in its totality.

If cosmology were just a branch of *applied mathematics* we could *define* it as the study of the global properties of "cosmological solutions" of certain field equations, notably Einstein's (cf. Hawking and Ellis 1973). We would then include singularities (such as the Big Bang) as boundary points of the Riemannian manifold modeling the universe. The qualification of an exact solution as a *cosmological* model still would have to be made by borrowing ideas from physics; for example, the kind of isometry group to be assumed. Consequently, homogeneous and isotropic cosmology models with *compact* space sections of *negative* curvature will have to be discarded because they admit only a 3-parameter isometry group, globally (Ellis 1971)[2]. The cosmological models of applied mathematics which, by careless use of language sometimes are called "cosmologies" (Ellis 1991, Halliwell 1991, Ryan and Shepley 1975) or "universes" (Salvati 1986, Robertson and Noonan 1968), need not have any relation to the external world.

A further *linguistic* ambiguity results from the use of the expression "cosmologies" in the sense of *alternative* theories (to Einstein's) for constructing cosmological models (Wesson 1978).

---

[2] Compare, however, a cosmological model with multiply connected space sections (Ellis and Schreiber 1986)

2.2   Further Features Peculiar to Cosmology

Besides the peculiarity that we do not exactly know what the physical system "universe" is, there are other epistemological, methodological, ontological, and semantical specialities making me wary of accepting cosmology right away as part of physics.

One striking difference may be found in the meaning of the concept "prediction". In cosmology, without exception, prediction is a conclusion from *past* times to the *present*. Slightly changing a statement of Walter Benjamin, cosmologists are prophets for the past. In physics proper, prediction means the foretelling of a *future* state from conditions given *now*. The social usefulness of natural science (and technology) rests on this regular meaning of prediction. Certainly, cosmological models can be used also to make correct calculations toward the future (Dyson 1979, Discuss et al. 1985). These calculations are pointless, however, because they cannot be checked by measurements on the relevant cosmological timescales (Will any of these calculations be conserved for 10^6–10^{10} years?). Even if cosmological theory could provide us with a reliable description of the past, its validity for the future is not guaranteed. If the concept "prediction" is used nonchalantly in the sense of, for example, "Einstein's equations *predicting* graviational waves" then only a semantical problem is left over.

Another characteristic feature of the universe is its *uniqueness*: one and only one such physical system can be thought of as existing. Leibniz' "best of all possible worlds" was the sole *actual* universe from an infinity of *thinkable* models. Unfortunately, with the advent of quantum cosmology a semantical erosion of the word "universe" has begun. We are asked "How many universes are there?" and authors investigate "a dilute gas of universes" or a "single parent universe ... in a plasma of baby universes" (Strominger 1991). A "Fock space of universes" is also employed (Fischler et al. 1989). We are seriously invited to "suppose universes are emitted from $t = 0$ like photons from an antenna" (Süsskind 1991). One wonders, though, what kind of tangible receptacle (a "superuniverse"?) could house or receive universes. Not to speak of the fact that, in our current understanding, concepts like space, time, matter, and the universe are inextricably linked.

Possibly, cosmology as the science of the physics of the universe is being transformed into a science of mathematical theories for the construction of cosmological models. In this case, a "dilute gas of universes" perhaps could be reformulated as "a sparse set of (pre-)cosmological models".

The many-worlds interpretation of quantum mechanics, an outlandish aid for the interpretation of the quantum mechanical measuring process, seems to have opened a flood gate. Tipler defines the Universe to consist of all logically possible universes where "Universe" is the totality of everything in existence and "universe" a single Everett branch (Tipler 1986, Tipler and Barrow 1987). In my view, he surpasses the simpleminded definition of the universe as a very large system of interacting masses by including in it thought processes as essential ingredients[3].

If the uniqueness of the universe is accepted, why then is it so special? Isn't the Earth unique, too? True, as far as its *individuality* is concerned. But the Earth is just one of nine planets in the solar system and one of billions more conjectured around other stars. It gets its individuality *by comparison* with other planets. Is there an empirical or a conceptual way of comparing "our" universe to "others"?[4] There is also a difference between the universe as it exists (if it exists) and, say, a single ion in an ion trap. Such a single system can be prepared from an ensemble of ions and the *ensemble* average may be replaced by a *temporal* average. For the evolving universe, only a temporal average can be realized, in principle.

As a consequence of the uniqueness of the universe *specific cosmic laws cannot exist* (Munitz 1963). It is not excluded that new physical laws will be discovered while we try to scientifically describe the cosmos. Such laws, however, must refer to properties of parts (subsystems) of the universe and to relations among them.

Can theories applying to a single object be falsified? The example of the *steady-state* cosmological model shows that falsification is possible for statements of cosmological theory pertaining to the past, because observations made now are observations of past states of the universe. In this regard cosmology is similar to *historical* science; falsification means nothing more than that our interpretation of the historical records has been wrong.

Since the Einstein field equations for the cosmological model are *hyperbolic* partial differential equations, a *Cauchy initial value problem* with given initial data must be solved in order that we may arrive at a unique solution. An additional chain of argumentation or even a theory must be developed by which the initial data actually in effect for the universe as we observe it are picked out from among the possible initial data. Thus *cosmogony*, the theory of what brought the cosmos into being, and cosmology are inseparable Siamese twins[5].

In fact, a lot of recent interest within cosmology is focused on the search for a rationale for the particular initial data required for the universe to be as it appears to be. Already within classical theory, attempts in this directions have been made (Collins and Hawking 1973, Misner 1968). Also, the *anthropic*

---

[3] Tipler might protest this formulation, however, because, to him, a logically possible universe and the actual universe seem to be of the same ontological quality.

[4] In contradistinction, *cosmological models* can be compared with each other – on paper, though.

[5] The assumption of temporal closure of the universe is the only escape in sight. With its painful consequences for causality and pre(retro-)dictability, the idea has not yet been taken seriously.

principle (Carter 1974) has been invoked[6]. The rise of quantum cosmology expresses the most massive attempt to bring cosmogony into the reach of science (cf. Sect. 3.3).

As a remark on the side: a related question is whether observation of the physical system "universe" will permit us, in the near future, to reconstruct its initial state. Even for as simple a system as the solar system the task is rather difficult. From what can be learned from deterministic chaos, and in view of the possibility that the Einstein field equations need not be a lasting foundation of cosmophysics, we should remain reserved in this matter.

Now, with the particular traits described shouldn't we ascribe *individuality* to the cosmos (Munitz 1986)? In this way, the concept "universe" would appear even closer to the understanding of the concept "God" in the Judeo-Christian tradition. My answer is negative: individuation cannot occur for a singly existing entity.

2.3   Cosmological Questionnaire

To get back to more solid ground we make a list of questions to be answered by cosmology:

1. Is *space* of finite or infinite extension?
2. Is *time* of finite or infinite duration
   (a) in the future,
   (b) in the past?
3. Is there a cosmic dynamics?

If the system is finite in space and in past time we may ask for the *total mass* (momentum, angular momentum) and the *age of the universe*. In case there is a dynamics, the initial state of the universe and its evolution in time are of interest. Further questions will arise within the three phases of cosmological modeling to be discussed now. If cosmology is a science, it is hoped that such questions can be answered in the future (in some generations?).

3.   Cosmological Modeling

3.1   The Standard Model of the Universe

3.1.1   General Hypotheses

As far as the universe is traced by its large scale mass structures (galaxies, clusters of galaxies, superclusters), the questions asked are concerned with the angular and in-depth distribution of such structures, their material content, the occurrence of the chemical elements, the origin of particular objects, e.g.,

---

[6] In my opinion, the anthropic principle (in its various forms) is nothing but a demand for selfc-onsistency of the cosmological model. Its explanatory power is null, in cosmology.

quasars, the strength and time-evolution of cosmic radiation fields, etc. From the answers, *average* properties will be ascribed to the universe serving as entries for cosmological model building.

Before a *quantitative* description of the universe can be achieved, a number of fundamental assumptions must be made. We list a few:

A_1 The physical laws, in the form in which they are valid here and now, are valid *everywhere* and *for all times.*

A_2 The values ascribed to the fundamental constants here and now are the same everywhere and at all times.

Tacitly assumed is always a *principle of simplicity* demanding that the simpler cosmological model is the better one. The following requirements are reflections of this principle:

A_3 The universe is connected

A_4 The material substrate of the universe is described by an *ideal* gas (fluid) – not a viscous fluid.

A_5 The material substrate of the universe evolves in time as a *laminar* flow – not a turbulent one.

Such hypotheses should be testable by their consequences, in principle. Today, however, we are still far away from such empirical checks. Often, A_2 is considered to be well founded, *empirically* (Fang Li Zhi and Shu Xian 1989). In my opinion, this is not as certain as we might wish it to be. The value of $\dot{G}/G \lesssim 10^{-11}$ (G: Newtonian gravitational constant) follows from measurements in the solar system covering the past 200–300 years (Will 1981) or from the arrival times of signals from the binary pulsar PSR 1913 + 16 measured since 1974 (Damour et al. 1988). At best, the observation time could be extended to $\sim 10^9$ years, i.e., the lifetime of the solar system. This is short if compared to the Hubble time. Thus, the conclusion concerning $\dot{G}/G$ drawn from direct observation are strongly dependent on theory. The situation is worse for the estimates of $\dot{G}/G$ made from primordial nucleosynthesis (Accetta et al. 1990) and giving a value for the ratio of G taken at the time of big bang nucleosynthesis and at present. Again, except for ^{4}He, the *observed* distribution of the light elements comes from measurements within the solar system only.

The constancy of the electron-proton mass ratio and the weak and strong interaction coupling constants rests on comparably weak empirical checks taken from redshift data out to $z = 2.7$ (formation of galaxies is supposed to occur for $z \simeq 10$–30) or the radioactive decay of elements which, in our present understanding, could have formed only in stars (Shlyakhter 1976, Rozenthal 1980, Cohen 1988). Star formation, however, does not occur within the first quarter of the assumed age of the universe. The time-independence of the fundamental constants, which is particularly important in the second phase of cosmological modeling, the early universe, is not directly verified during this period. All that can safely be said today is that A_1 and A_2 are not in conflict with the empirical data available. A_1 expresses the hope that local physics and the physics of the universe are *not* inextricably interwoven: "physics on a small scale determines physics on a large scale" (Ohanian 1976). The opposite view, i.e., that "the physical laws, as we usually state them, already involve the universe as a whole" is considered fundamental, occasionally (Hoyle and Narlikar 1974).

In addition to fundamental suppositions for *theoretical* modeling, hypotheses for the gaining of data and the empirical testing of cosmological models are necessary. Such are, for example:

B_1 The volume (spatial, angular) covered by present observation is a *typical* volume of the universe.

B_2 Observation time is long enough to guarantee reliable data of cosmological relevance.

It is doubtful whether these demands are satisfied at present. Three-dimensional redshift surveys of galaxies comprise a few thousand galaxies within a large angular region and with redshifts $z \lesssim 0.03$; two-dimensional surveys of a narrow strip of the sky have $z \lesssim 0.05$. There are also deep one-dimensional surveys ($z \lesssim 0.5$) covering a field of $1^0 \times 1^0$ with a few hundred galaxies (Broadhurst et al. 1990). In view of a supposed total of 10^{11} galaxies and the fact that angular position surveys extend only to depths of $\simeq 10\%$ of the Hubble length, one cannot say that these surveys are exhaustive. The same is true for an observation time of less than 10^2 y relative to an assumed age for the universe of $\simeq 10^{10}$ y.

Both A_3 and B_1 are questionable also due to the existence of *horizons* in many of the cosmological models used. There may be parts of the universe not yet observable (*particle* horizons) or parts which, in principle, cannot be observed from our position, i.e., event horizons. The problems related to observations were investigated carefully by G. Ellis (Ellis 1984)[7]

3.1.2   The Standard Model

In the standard model, the lumpiness of matter in the form of galaxies or clusters of galaxies is neglected in favor of a continuum model of smeared-out freely falling matter. The equations of state considered refer to pressureless matter (baryon dominated universe) or to radiation where $p = 1/3\mu$ (radiation dominated universe)[8] Sometimes, a combined 2-fluid model is used.

The gravitational field and space-time are described by a Riemannian manifold with Lorentz metric. Freely falling gas particles follow timelike geodesics of this metric. An important hypothesis for the standard model is the *cosmological principle:*

---

[7] In the very early cosmos, the problems stemming from the existence of horizons, possibly, are lightened by the application of quantum field theory (Wald 1986).

[8] Here, p is the pressure and μ the energy density of the ideal gas. For a detailed discussion of the standard model (and the early universe) see (Boerner 1988).

A_6 No matter particle has a preferred position in the universe.

Consequently, the space sections of the spacetime manifold describing the universe are homogeneous and isotropic. The cosmological model is given by a Friedman-Lemaître solution of Einstein's field equations (with or without cosmological constant). As the main result we find *qualitative compatibility* with observations of cosmological significance: the *expansion of the universe* (redshift) and the *isotropy of the slices of equal time* (cosmic microwave background). The isotropy does not refer to the position of the earth, the solar system or the galaxy but to an imagined rest system of smeared-out matter.

From the observational point of view homogeneity of the space sections is a fiction. The scale of homogeneity for which averaging of the large scale observed inhomogeneities (superclusters, voids) is reasonable has steadily increased in the past and could grow further in the future. From observations alone, it seems impossible to discriminate between a Friedman model and a spatially inhomogeneous, *static* model resembling, in our neighborhood, a Friedman model (Ellis et al. 1978). Friedman's model wins out, because the observed redshifts are interpreted as marks for the expansion of the universe. It is possible to derive, in theory, homogeneity from isotropy plus other plausible assumptions (Ehlers et al. 1968).

The Friedman cosmological model does not care whether its primordial states are warm or cold. Only after we add (non-relativistic) thermodynamics by assuming that the expansion of the universe is an *adiabatic* process can the decline of temperature with the extension of space be shown. In consequence it is possible to interpret the 2.7 K microwave background as a relic of an early, hot phase of the universe. On the other hand, the hypothesis of adiabaticity might be in conflict with the second law of thermodynamics if applicable to the universe. From local physical processes we expect the entropy of the universe to grow with the expansion (evolution of objects).

Mathematically, the most important consequence of the Friedman model is that it shows the existence of a density – as well as a metrical – *singularity* occurring in the *finite* past, the famous big bang. By the theorems of Penrose and Hawking (Hawking and Ellis 1973), this singularity receives a *generic* significance within cosmological model building. Nevertheless, from the point of view of physics, the infinities inherent in the big bang cannot be taken seriously.

As we have seen in Sect. 2.2, the predictive power of the standard cosmological model is nothing more than an expression of self-consistency: if the temperature at one *past* time, i.e., at the decoupling of radiation and matter, was such and such, then we should measure microwave *and* neutrino backgrounds of temperatures 2.7 and 1.9 K, respectively. If the backgrounds observed were at other temperatures, the initial data would have to be changed. If they were not observed at all, some part of the modeling (e.g., the application of non-relativistic thermodynamics) ought to be replaced by a better idea.

The standard cosmological model must face the problem of how to get away from the homogeneity and isotropy of the averaged-out large scale matter content in order to arrive at an explanation of the galaxies, clusters of galaxies etc. The hypothesis of adiabatic Gaussian density fluctuations with a scale-invariant spectrum together with such various competing scenarios as *cold* or *hot dark matter* (in the form of weakly interacting particles), *cold baryon matter*, cosmic string perturbations, local explosions, etc., is not consistent with the full range of extragalactic phenomena (Silk 1987, Peebles and Silk 1990).

Modeling in the first phase of cosmology runs parallel to modeling in other parts of physics. We give a condensed scheme in Fig. 1. Coming back to the questions asked in Sect. 2.3, it is still impossible to answer whether space is of finite or infinite extension and whether time is of finite or infinite duration in the future.

3.1.3   Philosophical Problems

The standard model of cosmology is not free of epistemological and method-ological problems. In it, Newton's absolute space appears in disguise in the form of an *absolute reference system*. In particular, (absolute) *cosmic time* or epoch is without *operational* background: the only clock showing it is the universe itself. *By definition*, cosmic time is identified with atomic time. By what sequence of clocks, the measured time intervals of which are overlapping, can we cover a duration of $2 \cdot 10^{10}$ y and more?

Similarly, there is *no operational* way of introducing simultaneity. The local method of signaling with light cannot be carried out, in practice, if distances of millions of light years are involved and the geometry in between is uncertain. It cannot be used, *in principle*, for the full volume of space if event horizons are present.

The cosmological models containing the concept of "simultaneous being of part of the universe" (technically, the space sections or 3-spaces of equal times) are catering to past pre-relativistic needs. In the relativistic space-time concept, access to the universe is gained through the totality of events on and within our past light cone. Hence, "simultaneous being" is replaced by "what may be experienced at an instant at one place". Some of the objects at the sky, the radiation of which we observe now, may not exist anymore. Observational cosmology can be compared to an archeological excavation: deeper and deeper strata of the past are uncovered, with the difference to real archeology being that the present state of the objects found cannot be known empirically. The universe is not a museum, it is a dynamical system.

A fundamental hypothesis going into the standard model is the concept of a time *common* to all parts of the universe. In some cosmological model (for example, Gödel's), the local spaces of simultaneity are not integrable to one and only one 3-space of "simultaneous being". A much more drastic approach would use a relativistic *particle* model (in place of the present *field* model),

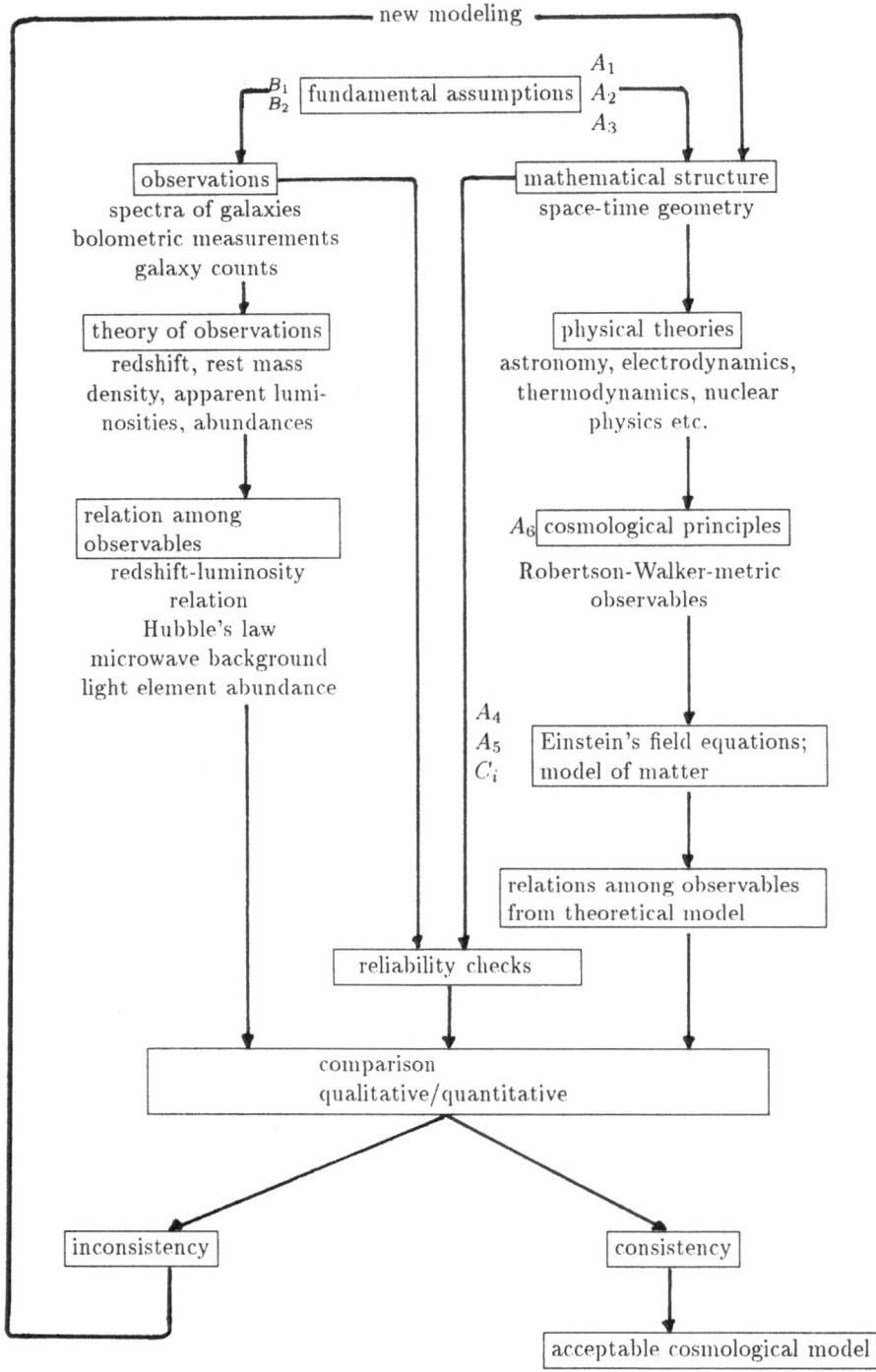

Fig. 1. Scheme for cosmological modeling within standard cosmological model

in which each particle (each part of the universe) has its own time. Such a model is possible, in principle, but has not yet been studied in detail due to its complexity.

3.2 The Early Universe

3.2.1 Particle Astrophysics

The second phase of cosmological modeling is characterized by a change of paradigm: while, in principle, the description of matter by a continuous distribution in the form of an ideal gas is retained, in practice the cosmological substrate now is no longer dealt with collectively. Matter is differentiated into elementary units: atoms, nuclei, elementary particles. They interact, can be produced and split or annihilated. The interplay of elementary particle reaction rates and the expansion rate of the universe leads to different equations of state for different particle species at the same epoch. This phase of cosmological modeling applies to the time interval between the Planck time and $\simeq 10^2$ seconds (the end of primordial nucleosynthesis) after the big bang. The 47 powers of 10 covering this period (in seconds) are to be compared to the only 15 powers of 10 in time from primordial nucleosynthesis until now.

Empirical access to the early universe is gained by the observation of the abundance of chemical elements, say of ^{4}He, and the impossibility of reproducing it, quantitatively, by thermonuclear processes *in stars*. If the 2.7 K microwave background is interpreted as a relic of the early, hot universe, the Friedman solutions may also be extrapolated to such early times. It then seems permissible to transfer our knowledge of nuclear reactions from the laboratory to the early states of the universe. Nucleosynthesis for the light elements d, ^{3}He, ^{7}Li, except for ^{4}He, depends sensitively on the single parameter of cosmological relevance entering: the ratio $\eta = n_B/n\gamma$ of the number of baryons to the number of photons in the universe. $n\gamma$ can be calculated from the microwave background. The decisive nuclear physics parameter is the neutron's lifetime. Nucleosynthesis calculations lead to a value for the average matter (baryon) density of the universe *consistent* with what is observed, directly, from luminous masses and, indirectly, through dynamical effects in galaxies and clusters of galaxies. Because the production of ^{4}He is dependent on the number of existing neutrino families, it is possible to obtain an estimate consistent with what has been found with the largest particle acclerators (Walker et al. 1991).

Again, cosmological modeling of the early universe is based on a number of hypotheses, a selection of which follows:

C_1 Elementary particles are pointlike.

C_2 Nuclear and elementary particle reactions and reaction rates in the early universe are the same as found today, in the laboratory.

C_3 The elementary particles do not interact gravitationally; gravitation acts merely as an external field.

C_4 While, in each epoch, matter is in thermodynamic equilibrium, different particle species can and will decouple from the equilibrium distribution.

C_5 The temperature of the universe can grow without limit for decreasing epoch.

With these assumptions the modeling of the early, hot states of the universe is simplified. As to C_1, obviously, it is difficult to imagine how $\simeq 10^{80}$ baryons could find room in a universe thought to be of a linear dimension $\sim 10^{-33}$ cm. C_2 might be read as a subcase of A_1. There is more to it, however, because as long as we cannot *control* thermonuclear reactions, the conditions prevailing in the early universe cannot be reproduced in the laboratory. We might just not yet know all the physical laws needed.

Surprisingly, gravitation plays a subordinate role in the modeling of the early universe despite the belief that matter was then extremely condensed. The gravitational field shows up only in the expansion of the universe or, perhaps, in pair production of elementary particles if quantum field theory in curved space as we understand it is applicable.

As to thermodynamics and kinetic theory, the coupled Einstein-Boltzmann equations are not solved. What is solved, *approximately,* is the Boltzmann equation in a *prescribed* gravitational field. It is known that, in the Friedman cosmological models, an equilibrium distribution is permitted only in two limiting cases: the ideal radiative model (rest mass of particles is zero) and the "heavy mass" model (rest mass is infinite) (Bernstein 1988). Thermodynamically, the expanding universe is treated as a *quasistatic* system. Otherwise, the time dependence of the cosmic temperature following from the cosmological model would have to be interpreted as a characteristic sign of a *non-equilibrium* system. The concepts temperature and entropy of the universe can be defined through kinetic theory (statistical mechanics) only: no "external" heat bath is available.

Finally, C_5 restricts possible equations of state such as Hagedorn's (Hagedorn 1965, 1973)[9] leading to a maximal temperature $k\,T_{\max} \simeq m_\pi c^2$ (m_π : mass of the pion) which is $\simeq (1-2) \cdot 10^{12}$ K.

Prior to the epochs in which primordial nucleosynthesis occurs, the relevant energy scales get much higher than those that can be probed, at present, on Earth. It is believed that, with growing energy, the different fundamental interactions are united or, vice versa, that there is spontaneous symmetry breaking in the course of the time development of the universe.

---

[9] Zeldovich and Novikov exclude Hagedorn's equation of state by arguments concerning vacuum polarization (Zeldovich and Novikov 1983).

3.2.2   The Inflationary Model

If the validity of the Friedman cosmological models is extrapolated to early epochs ($\leq 10^{-30}$ seconds after the big bang), a number of questions arise:

> What makes the universe as isotropic and homogeneous as it is? (horizon problem)
> Why does the density parameter Ω_{mat} differ from Ω_{crit} by only two factors of ten?[10] (flatness problem)
> How can the ratio $\eta = \frac{\eta_B}{\eta_B\gamma} \simeq (4-7)\cdot 10^{10}$ be explained? (entropy problem)

In order to answer these questions the *inflationary model* was invented (Guth 1981, Albrecht and Steinhardt 1982, Linde 1990, Kolb and Turner 1990). Its characteristic feature is the introduction of a scalar field ϕ which is supposed to dominate the matter content at very early epochs, but is quickly radiated away[11]. Usually, although not necessarily, ϕ is taken to be the order parameter of a phase transition from a symmetric phase with high energy corresponding to $\phi = 0$ to a phase with *broken symmetry* and $\phi = \text{const} \neq 0$. An analogue would be the delayed transition from the gaseous to the fluid state with undercooling. The phase transition is made to start at $\simeq 10^{-35}$ seconds after the big bang. Dynamically, it is tripartite: after the tunneling of a potential barrier between the false and the true vacuum, the slow descent toward the true vacuum occurs (supercooling) passing over into a brief period (as compared to the Hubble time) of field oscillations (reheating). In this last interval, the energy of the false vacuum, which in Einstein's equations shows up in the form of a cosmological constant, is radiated away.

The phase transition is pictured as a nucleation of bubbles of the broken-symmetry phase within a matrix of the symmetric phase. During supercooling such a bubble can grow exponentially by 40–50 orders of magnitude (of 10) and more within a time of the order of a (few hundred) $\cdot 10^{-35}$ seconds. The gravitational field during the exponential growth is described by DeSitter's solution of the field equations, the space sections of which are *flat*. The re-heating process is non-adiabatic, bringing an increase in the entropy (of the universe) by a factor of 10^{130}.

The inflationary model thus can solve the entropy problem as well as the horizon problem: the presently observable part of the universe lies within a single inflating bubble, which means that, at the epoch of decoupling of photons and baryons, the various regions of the universe from which the cosmic microwave background originates could have been causally connected. The model is also said to solve the flatness problem: inflation drives the density parameter Ω toward one (Ellis 1991). Whether $\Omega = 1$ is desirable or not seems to be entirely up to one's private beliefs. $\Omega = 1$ is an unstable fixpoint in the

---

[10] $\Omega_{mat} := \mu/\mu_{crit}$ where μ is matter energy density and $\mu_{crit} := \frac{3H_0^2}{8\pi G c^4}$ with H_0 being the Hubble constant. $\Omega_{mat} = 1$ leads to a cosmological model with *flat* space sections.

[11] The field quantum of this scalar field sometimes is called the "inflaton".

phase diagram of the time evolution of the Friedman models. The value $\Omega = 1$ prescribed to the universe seems off the mark. If adopted, it would imply that 90 % of the universe's matter is dark and, possibly, nonbaryonic.

If we accept that the questions asked above require an urgent answer and that the inflationary model provides a satisfactory response, other questions arise. Where does the mysterious scalar field ϕ come from? It is not the Higgs field (which, by the way, is also not yet observed). Is it connected to a field theory describing spontaneous symmetry breaking of supersymmetry? Will there be a *technically accomplished* model for inflation still lacking now? One related problem is that the scalar field must be very weakly coupled to all other matter fields. What determines the initial value of ϕ, i.e., the high energy of the false vacuum? Can we *observe* traces of the inflationary period of the very early universe? Other effects following from the model such as a stochastic background of gravitational waves (Grishchuk 1975) and a scale invariant spectrum for density perturbations have not yet been tested empirically. Also, the quantum fluctuations of the scalar field are calculated in the *prescribed* DeSitter metric. In view of the many open questions, the *explanatory value* of the inflationary model is debatable.

3.3   Quantum Cosmology

3.3.1   Law of Inertial Conditions?

In the third phase of cosmological modeling, the epoch *around and before the Planck time* is dealt with. The general understanding is that, at such extremely early epochs, *quantum mechanics* must be applied. Consequently, both the gravitational field and all matter fields are to be quantized. Despite the fact that, at present, an acceptable quantum field theory of gravitation, *quantum gravity*, is not in sight, various incomplete schemes, for example canonical quantization (DeWitt 1967, Wheeler 1968), invented for the quantization of the gravitational field, are applied. The approach rests on three hypotheses:

D_1 The gravitational field must be quantized at and before the Planck epoch.

D_2 The quantization procedure, in principle, does not differ from field quantization of the other fundamental interactions.

D_3 By standard canonical quantization the essential degrees of freedom of the gravitational field are quantized.

D_1 is the majority vote. A minority believes in gravity as a *classical* field generated, perhaps, as an *effective* field through the other fundamental interactions. D_2 is at the root of the difficulties of quantum gravity: it has not yet been possible to implement this hypothesis in a convincing manner. In particular, it still is not very clear whether it suffices to quantize the gravitational field on a continuous space-time or whether the very concept of a manifold ought to be replaced by a discrete set (lattice, etc.). Competing with D_3 would be an approach using Ashtekar-variables (Ashtekhar 1991). Up to now, the dynamical degrees of freedom have not been identified unambiguously.

Surprisingly, the program of applying quantum mechanics to the universe is *not* seen as an *intermediate* step in between the big bang and, say, the inflationary epoch with the aim of providing the initial conditions required for inflation. It is taken as a program for a *cosmogonic* theory: an attempt to construct a theory *determining uniquely* the initial conditions of the universe (Hartle and Hawking 1983, Vilenkin 1988, Gell-Mann and Hartle 1990)[12].

Such an endeavor makes sense only if the universe itself carries the rationale for its initial data. Otherwise, we should have to take recourse to some sort of superuniverse or a creator. If transferred to human life the idea would mean that the reason for my coming to life does not lie in my parents but is in myself. Strange, even absurd as this thought may be, a human being and the universe are quite different systems. It seems plausible that, ontologically, the cosmos cannot be thought of without the inclusion of a reason for its existence. It is unnecessary to add that the very idea of *prescribing uniquely* the initial data of a system with the help of its dynamics violates the spirit of present-day physics. Nevertheless, quantum cosmology is said to provide us with a law of initial conditions (Halliwell 1991).

3.3.2   The Wheeler-DeWitt Equation

If quantum theory is applied to the universe, canonical quantization leads to the Wheeler-DeWitt equation, an analogue of the stationary Schrödinger equation for the *wave function* of the universe ψ. It is a functional $\psi[^3g, \phi]$ of the geometry of space sections and matterfields ϕ and hence defined on an infinite-dimensional space called *superspace*. The spacetime geometry can be pictured as a trajectory in superspace. The wave function of the universe represents the superposition of all possible space-time geometries correlated with matter functions (Zeh 1986).

In practice, the infinite-dimensional superspace is reduced to a finite number of degrees of freedom, i.e., to *minisuperspace*. Very often, in model calculations, isotropy and homogeneity of the space geometry is assumed and leads to a wave function ψ depending on just one geometric variable: the scale factor a of the Friedman models. Moreover, only a single scalar matter field ϕ is taken into account such that $\psi = \psi[a, \phi]$. Despite this technical simplification, the main problem cannot be circumnavigated: a *unique* solution of the Wheeler-DeWitt equation is obtained only if a *boundary* condition for ψ is chosen. Several suggestions to this end have been made and discussed. In the path integral formulation (Hartle and Hawking 1983, Hawking 1984) ψ is determined by summing over all paths describing *compact* (euclidean) 4-geometries with regular matter fields. All 4-geometries must have a given 3-geometry as their boundary (no-boundary-condition)[13]. The resulting wave

---

[12]I cannot take seriously a use of language in which "cosmologists hope to look beyond the very instant of creation" (Halliwell 1991).

[13]As C.J. Isham (Isham 1987) says, "the universe is created ex nihilo since the 4-manifold has only the connected 3-space as its boundary".

function, however, is not normalizable, in general. An alternative condition is Vilenkin's quantum tunneling from nothing (where "nothing" corresponds to the vanishing of the scale factor a): the universe is nucleating spontaneously as a DeSitter space (Vilenkin 1982, 1984, 1988). This boundary condition has been criticized on the ground that it equally well describes tunneling *into* nothing.

By now, it should have become clear that the wave function of the universe does not depend on an *external* time parameter. In minisuperspace, the Wheeler-DeWitt equation is a *hyperbolic* differential equation the dynamics of which depends on two variables, a and ϕ, both of which can play the role of an *internal* time. The ambiguity in the selection of an internal time parameter permits reinterpretation of the Wheeler-DeWitt equation as a Klein-Gordon equation. This is used by another version of quantum cosmology: *third quantization* of general relativity. Here, change of topology of 3-geometry plays an important role (Strominger 1991). Now, the universe is the field quantum of the Wheeler-DeWitt equation and thus universes are created and anihilated as if they were particles. No wonder then that graphs similar to Feynman graphs are used to depict the various "generations" of "universes". The vacuum, so it seems, is on the verge of becoming the *omnipotent agent:* not only does it house virtual particles of all sorts, but virtual "universes" as well.

3.3.3   Puzzles of Quantum Cosmology

Obviously, the Copenhagen interpretation of quantum mechanics cannot be applied to the wave function of the universe. Otherwise, one would have to explain with what kind of measuring instrument a state of the universe could be prepared and measurements of observables be made. Only the universe itself might be a proper measuring apparatus. Who is the classical observer carrying out preparation and other measurements? Perhaps most of the universe is measuring the remainder (Finkelstein and Rodriguez 1986)? A continuous shift of the borderline between observing and observed parts of the universe would then be necessary.

It seems unavoidable to employ some version of Everett's interpretation of quantum mechanics wherein the splitting of the wave function by a measurement is equivalent to splitting the universe into many possible ones. In each of these universes one of the allowed measurement results would occur (Everett 1957). In a recent development the "many worlds" of Everett are replaced by the "many histories" interpretation in which observers making measurements are within "decohering" histories of the same universe (Gell-Mann and Hartle 1990). A related problem is the derivation of *classical* properties of the universe from the wave function in quantum cosmology (Zeh 1971, Joos and Zeh 1985): classical properties are expected to emerge by a continuous measurement process (Kiefer 1988).

How does *cosmological time* emerge from the Wheeler-DeWitt equation – the 3-geometry being postulated *ab initio*?[14] Presently, at best a perturbative approach might lead from internal to external time, if it is possible at all to make the transition. A further problem refers to the *arrow of time*. Is it possible to choose the initial data of the universe such that the direction of time is explainable? Roger Penrose suggests assuming homogeneity of space – corresponding to a low value of entropy – as an initial condition. The Weyl tensor is tentatively used as a measure of the universe's entropy (Penrose 1979, 1986, 1989) and required to vanish at singularities in the past. In such an approach, the hypothesis

D_4 Einstein's field equations hold right up to the big bang singularity

is implied.

4. The Science of Cosmology

We have seen that cosmology shows features of descriptive astronomy, palaeontology, history, mathematics, physics, and natural philosophy. As long as cosmology is handled as *cosmophysics*, i.e., as an extension of physics from the galactic through the extragalactic realm to ever larger massive gravitating structures, it is part and parcel of physics proper. In comparing the standard cosmological model with the model of plate tectonics in the Earth sciences (Fig. 2), we notice, apart from all parallelism, a scarceness of empirical data in its support. The situation becomes much worse for the modeling of the early universe.

4.1 The Explanatory Value of Cosmology

To me, the explanatory power of the standard model is weak. I find it difficult to understand why an extreme *global* thinning of matter against *local* gravitational attraction is needed first while, then, the massive superstructures arise from *local* condensations against *global* expansion. Not to speak of the distastefully aggressive beginning of the universe: the big bang. I also have problems accepting the empirical basis for one of the two pillars of the standard cosmological model: the abundance of chemical elements. Up to three different mechanisms are necessary to reach – more or less – quantitative agreement: primordial nucleosynthesis, nucleosynthesis in stars, spallation processes in the interstellar or intergalactic medium. Also, the comparison of calculated and observed abundances depends highly on theory (models for the chemical evolution of galaxies and stars).

---

[14]The inequality in the treatment of time and space is striking. Also in this regard quantum cosmology fails to be a convincing *cosmogonic* theory.

	cosmology	earth sciences
model	homogeneous-isotropic standard model	plate tectonics (6 shiftable plates)
physical theory	Einstein's theory of gravitation	largescale convective flow in the Earth's mantle. Energy furnished by radioactive decay (U,Th,K)
observations	2.7 K microwave radiation abundance of D,^{3}He ^{4}He, ^{7}Li	magnetization of rocks morphology of continents ocean floor geology paleontological findings
acceptance by specialists	high	high
reasons for acceptance	cosmological model gives a better empirical basis for both general relativity and high energy physics; interdisciplinary work	explanatory and predictive fruitfulness (prediction of earthquakes?) interdisciplinary work
reasons for refutal	empirical basis is too weak; lack of confidence in extrapolation of physical laws	?

Fig. 2. Comparison of standard cosmological model and model of plate tectonics

It is difficult, from the theoretical point of view, to make transparent the web of assumptions, logical deductions, and empirical input spun by cosmologists, particularly if we wish to evaluate the explanatory and predictive value of cosmology. Special case studies seem to be called for. Hypotheses of differing weight are intermingled, for example, the classical, *relativistic, nonlinear* theory of gravitation, *nonrelativistic* thermodynamics and kinetic theory, the *linear* theory of density fluctuations, quantum field theory in curved space, nuclear and high energy physics, etc. Poorly defined concepts such as the entropy of the universe, thermodynamical equilibrium of the universe, wave function of the universe are used without hesitation, except perhaps for vague apologies for unsettled conceptual problems involved.

Authors praise their "courage and boldness in attacking problems once thought to be beyond the reach of human comprehension" (Kolb and Turner

1991) while apparently believing that a tiny circle drawn on a piece of paper and annotated by "the presently observable Universe at the Planck time ...(100 × magnification)" gives a witty representation of the cosmos. This reminds me of Valéry's "l'univers n'existe que sur le papier". Weltanschauung and the prejudices so easily induced by it have recently invaded cosmology to an extent unknown to other subdisciplines of physics. The main drive seems to originate in a desire for reducing the multitude of appearances to a *single* cause, the esthetic dream of many physicists and natural philosophers. In that spirit, the realization of unified field theory or of the *genetic* concept of the cosmos, i.e., the belief that all relevant information about the universe is encased in its initial data, is pursued[15]. In order to reach such a goal, cosmological modeling is changed into a bird's view of the universe. Cosmologists seem to put themselves "outside" of the universe (mentally, that is) when trying to calculate probabilities of observing various possible configurations for the universe as it evolved until now.

4.2   Extrapolative Physics

It is time to name a new science *outside* of physics although intimately related with it. Let us call it *extrapolative physics* after its characteristic trait: an empirically uncontrolled extrapolation of physical laws. Quantum cosmology, the inflationary model, quantum gravity, black-hole thermodynamics, high-energy theories such as GUTs, SUSY-GUTs, supergravity, and string theory belong to extrapolative physics. Of course, all these theories could be classified as applied mathematics or mathematical physics as long as they did not insist on explaining something of the external world (nature) or part of it. Penrose calls such theories "tentative" (Penrose 1989). In connection with cosmology, we might even replace "theory" by "scenario" (Peebles 1991).

Extrapolative physics is a fecund interdisciplinary science taking place in our *heads* – as pure mathematics does. We also might call it *invention physics* because, by it, awareness of what is *potentially* real is produced. Passage from the potentially to the *actually* real requires the linking of extrapolative physics to an *empirical basis*. This can be achieved only by an agreement among scientists. Therefore, lengthy debates about cosmology's place are unavoidable[16].

---

[15] "At the moment we have a number of partial laws which govern the behaviour of the universe under all but the most extreme conditions. However, it seems likely that these laws are all part of some unified theory that we have yet to discover. We are making progress and there is a chance that we will discover it by the end of the century" (Hawking 1989).

[16] A rigorous classification might shut out from physics even the standard cosmological model. In Penrose's ranking the standard cosmological model is a "useful" but not a "superb" theory. Synge's statement that "of all branches of modern science, cosmological theory is the least disciplined by observation" (Synge 1966) would have to be shifted nowadays to string theory or supergravity, though.

It seems not undue, however, to demand of scientists doing pioneering work in the field of cosmology that they be aware of the fictitiousness of the "reality" they are producing. Reading through one or the other current *popular science* book on cosmology, I do get the impression that authors are easily carried away by the foolhardiness of their speculations.

To return to the question asked at the beginning: C. F. von Weizsäcker once compared the anti-mythical myth of the biblical story of creation with the mythical anti-myth of modern cosmology (v. Weizsäcker 1948). Silk calls cosmology a *falsifiable myth* (Silk 1987). If we (optimistically) believe that the level of rational thinking has risen since antiquity, we can, somewhat pungently, say that the science of cosmology – especially in dealing with the early and earliest epochs of the universe – is producing the cosmological myths adequate for our time.

References

Accetta, F.S., Krauss, L.M. and Romanelli, P. (1990): New Limits on the variability of G from big bang nucleosynthesis, Phys. Lett. **B 248**, 146-150

Albrecht, A. and Steinhardt, P.J. (1982): Reheating an inflationary universe. Phys. Rev. Lett **48**, 1437-40

Alfven, H. (1967): Kosmologie und Antimaterie, Umschau, Frankfurt

Ashtekar, A. (1991): Non-perturbative Canonical Gravity, World Scientific, Singapore

Barrow, J.D. and Tipler, F.J. (1986): The Anthropic Cosmological Principle, University Press, Oxford

Bernstein, J. (1988): Kinetic theory in the expanding universe, University Press, Cambridge

Boerner, G. (1988): The Early Universe – Facts and Fiction, Springer, Berlin

Bondi, H. (1961): Cosmology, 2nd Ed., University Press, Cambridge

Broadhurst, T.J., Ellis, R.S., Koo, D.C., and Szalay, A.S. (1990): Large-scale distribution of galaxies at the Galactic pole, Nature 343, 726-728

Buchdahl, H.A. (1981): Seventeen Simple Lectures on General Relativity Theory, John Wiley, New York

Cohen, E.R. (1988): Variability of the physical constants. In: Gravitational measurements, fundamental metrology and constants. Eds. V. DeSabbata and V.N. Melnikov, Kluwer, p. 91-103, Dordrecht

Collins, C.B. and Hawking, S.W. (1973): Why is the universe isotropic?, Astrophys. J. 180, 317-334

Damour, T., Gibbons, G.W., and Taylor, J.H.) (1988): Limits on the variability of G using binary pulsar data, Phys. Lett. **61**, 1151-1154

DeWitt, B.S. (1967): Quantum theory of gravity I. The canonical theory, Phys. Rev. 160, 1113-1148

Dictionnaire de la langue Française: Le Petit Robert (1985), Le Robert, Paris

Discuss, D.A., Letaw, J.R., Teplitz, D.C., and Teplitz, V.L. (1985): Die Zukunft des Universums, in Kosmologie, Struktur und Entwicklung des Universums, Spektrum der Wissenschaft S. 182-194, Heidelberg

Dyson, F.J. (1979): Time without end: Physics and biology in an open universe, Rev. Mod. Phys. 51, 447-460

Ehlers, J., Geren, P., and Sachs, R.K. (1968): Isotropic solutions of the Einstein-Liouville equation, J. Math. Phys. 9, 1344-1349

Ellis, G.F. (1971): Topology and Cosmology, GRG-Journal 2, 7-21

Ellis, G.F. (1984): Relativistic Cosmology, its nature, aims and problems. In: General Relativity and Gravitation. Eds. B. Bertotti, F. de Felice, and A. Pascolini, Reidel, pp. 215-288, Dordrecht

Ellis, G.F. (1991): Standard and inflationary cosmologies In : Gravitation. A. Banff Summer Institute. Eds. R. Mann and P. Wesson, World Scientific, p. 1-53, Singapore

Ellis, G.F., Maartens, R., and Nel, N. S. (1978): The expansion of the universe, Mon. Not. R. Astron. Soc. 154, 187

Ellis, G.F. and Schreiber, G. (1986): Observational and dynamical properties of small universes, Phys. Lett. A 115, 97-107

Everett, H. (1957): "Relative State" formulation of quantum mechanics, Rev. Mod. Phys. 29, 454-462

Fang Li Zhi and Ruffini, R. (1983): Basic concepts in relativistic astrophysics, World Scientific, p. 167, Singapore

Fang Li Zhi and Shu Xian (1989): Creation of the universe, World Scientific, p. 171, Singapore:

Finkelstein, D. and Rodriguez, (1986): Quantum time-space and gravity. In: Quantum concepts in space and time. Eds. Roger Penrose and Chris J. Isham, Clarendon Press p. 247-254, Oxford

Fishler, W. et al. (1989): Quantum mechanics of the GOOGOLPLEXUS, Nucl. Phys. B327, 157-177

Gell-Mann, M. and Hartle, J.B. (1990): Quantum mechanics in the light of quantum cosmology. In: Complexity, entropy, and the physics of information. SFI Studies in the Sciences of Complexity, Vol. 8. Ed. W.H. Zurek, Addison-Wesley, pp. 425-458, Redwood City

Grishchuk, L.P. (1975): Amplification of gravitational waves in an isotropic universe, Sov. Phys.-JETP 40, 409-415

Guth, A.H. (1981): Inflationary universe: a possible solution to the horizon and flatness problems, Phys. Rev. D 23, 347-356

Hagedorn, H. (1965): Statistical thermodynamics of strong interactions at high energies, Suppl. Nuov. Cim. 3, 147-186

Hagedorn, R. (1973): Thermodynamics of strong interactions. In: Cargèse lectures in physics, Vol. 6. Ed. E. Schatzmann, Gardon and Breach, p. 643-716, New York

Halliwell, J.J. (1991 a): Quantum Cosmology and the creation of the universe, Scientific American, Dec. 1991, 76-85

Halliwell, J.J. (1991 b): Introductary lectures on quantum cosmology. In Quantum cosmology and baby universes, Eds. S. Coleman et al., World Scientific, pp. 159-243, Singapore

Hartle, J.B. and Hawking, S.W. (1983): Wave function of the universe, Phys. Rev. D 28, 2960-2975

Hawking, S.W. (1984): The quantum state of the universe, Nucl. Phys. B 239, 257-276

Hawking, S.W. (1988): Eine kurze Geschichte der Zeit, Reinbek, Hamburg

Hawking, S.W. (1989): The edge of space time. In: The new physics, Ed. Paul Davies, University Press, p. 61-69, Cambridge

Hawking, S.W., and Ellis, G.F. (1973): The large scale structure of space-time, University Press, Cambridge

Isham, C. J. J. (1987): Quantum Gravity. In: General Relativity and Gravitation, Ed. M.A.H. Mac Callum, University Press, 114, Cambridge

Hoyle, F. and Narlikar, J.V. (1974): Action at a Distance in Physics and Cosmology

C. Kiefer, C. (1988): Wave packets in minisuperspace, Phys. Rev. D 38, 1761-1772

Kolb, E.W. and Turner, M.S. (1990): The early universe, Addison-Wesley. pp. 86, 498, Redwood City

Linde, A.D. (1990): Inflation and Quantum Cosmology, Academic Press, Boston

McVittie, G.C. (1961): Fact and Theory in Cosmology, Eyse and Spottiswoode, London

Misner, C.W. (1968): The isotropy of the universe, Astrophys. J. 151, 431-457.

Munitz, M.K. (1963): The logic of cosmology, Brit. J. Philos.Science 13, 34-50.

Munitz, M.K. (1986): Cosmic Understanding, University Press, Princeton

Ohanian, H.C. (1976): Gravitation and Spacetime, W.W. Norton, New York

Peebles, J.P.E. (1990): Cosmology: In: Encyclopedia of Physics, Eds. R.G. Lerner and G.L. Trigg, VCH, New York

Peebles, J.P.E., and Silk J. (1990): A cosmic book of phenomena, Nature 346, 233-239

Penrose, R. (1979): Singularities and time-asymmetry. In: General relativity, an Einstein Centenary, Eds. S.W. Hawking and W. Israel, University Press, pp. 581-638, Cambridge

Penrose, R. (1968): Gravity and state vector reduction. In: Quantum concepts in space and time, Eds. R. Penrose and C.J. Isham, Clarendon Press, Oxford

Penrose, R. (1989): The emperor's new mind, Oxford University Press, New York

Rozenthal, I.L. (1980): Physical laws and the numerical values of fundamental constants, Sov. Phys. Usp. 23, 296-305

Rowan-Robinson, M. (1977): Cosmology, Clarendon Press, Oxford

Sexl, R.U., Urbantke, K. (1983): Gravitation und Kosmologie, 2. Aufl., Bibliographisches Institut, Mannheim

Shlyahkter, A. I. (1976): Direct test of the constancy of fundamental nuclear constants, Nature 264, 340

Silk, J. (1987): Galaxy formation: confrontation with observations. In: Observational Cosmology, IAU Symposium 124, Eds. A. Hewitt et al., p. 391-413

Strominger, A. (1991): Baby universes. In: Quantum cosmology and baby universes, Eds. S. Coleman et al., World Scientific, p. 269-364, Singapore

Süsskind, L. (1991): Critique of Coleman's theory of the vanishing cosmological constant. In: Quantum cosmology and baby universes, Eds. S. Coleman et al., World Scientific, Singapore

Synge, J.L. (1966): Relativity Theory: The General Theory, North-Holland, Amsterdam

Tipler, F.J. (1986): The many-worlds interpretation of quantum mechanics in quantum cosmology. In: Quantum concepts in space and time, Eds. R. Penrose and C.J. Isham, Clarendom Press, p. 208, Oxford

Vilenkin, A. (1982): Creation of universes from nothing, Phys. Lett. 117 B, 25-28

Vilenkin, A (1986): Boundary conditions in quantum cosmology, Phys. Rev. D33, 3560-3569

Wald, R.M. (1984): General Relativity, University Press, Chicago

Walker, T.P., Steigmann, G., Schramm, D.N., Olive, A., and Kang, Ho-Shik (1991): Primordial Nucleosynthesis redux., Astrophys. J. 376, 51-69

Webster's Seventh New Collegiate Dictionary (1963): Merriam Company, p. 188, Springfield

Weizsäcker, C.F. von (1948): Die Geschichte der Natur, Hirzel, Zürich

Wheeler, J.A. (1968): Superspace and the nature of geometrodynamics. In: Batelle Rencontres, Eds. C. DeWitt and J.A. Wheeler, W.A. Benjamin, New York

Will, C.M. (1981): Theory and experiment in gravitational physics, University Press, p. 36-38, Cambridge

Zeh, D. (1986): Emergence of classical time from a universal wave function, Physics Lett. A 116, 9

Zel'dovich, Ya.B. and Novikov, I.D. (1983): Relativistic Astrophysics, Vol. 2, University Press, pp. 147, Chicago

Ziman, J. (1968): Public knowledge, University Press, Cambridge

II

Philosophical Concepts
and the Mathematics of Physics

On the Assumption That Our Concepts 'Structure the Material of Our Experience'

Felix Mühlhölzer

Seminar für Philosophie, Logik und Wissenschaftstheorie,
Ludwigstr. 31, D-80539 München 22, Germany

1. The Third Dogma of Empiricism

The general assumption that human knowledge depends, firstly, on our faculty to receive some raw material from the world outside us and, secondly, on our faculty to 'organize' or to 'structure' this material by means of certain concepts may sound quite innocent. At a close look, however, it appears to be anything but clear. In an article entitled "On the Very Idea of a Conceptual Scheme", the philosopher Donald Davidson has called this assumption "the third dogma of empiricism" (Davidson 1984, 189).

According to Willard Van Quine, the first dogma of empiricism is "the belief in some fundamental cleavage between truths which are analytic, or grounded in meanings independently of matters of fact, and truths which are synthetic, or grounded in fact", and the second dogma "is reductionism: the belief that each meaningful statement is equivalent to some logical construct upon terms which refer to immediate experience" (Quine 1961, 20). Quine has convincingly shown, I think, that these two dogmas – the dogma of the analytic-synthetic distinction and the dogma of phenomenological reductionism – are ill-founded. Davidson wants to show that the third dogma – the belief in a fundamental dualism of some material given to us and waiting to be organized, on the one hand, and some conceptual scheme doing the organization, on the other hand – is ill-founded as well. In what follows I shall pick up the thread of Davidson's reflections. I will try to work out, as clearly as possible, the questionable sides of this third dogma of empiricism and to replace it by a more adequate view.

The most emphatic expression of this dogma can be found in Kant. At the beginning of the *Transcendental Aesthetic* of his *Critique of Pure Reason*, Kant draws the distinction between the matter and the form of appearances. The matter is said to be that in appearances which is given in sensation; it is the neutral content of our experience. The form is said to be "that which brings it about that the manifold of appearance allows of being ordered in certain relations". This ordering is done by means of concepts contributed by the mind, or, more precisely, by our understanding. In order to have knowledge, both are needed, content and concepts, because "thoughts without content are empty, intuitions without concepts are blind".

Here we see the dogma of a fundamental dualism of empirical content and conceptual scheme in its most explicit form. Kant's philosophy, moreover, reveals features of this dogma which are by no means innocent. Since the conceptual structuring is done by the human mind, Kant detects in it certain aprioristic elements which are said to be necessary for empirical knowledge. These aprioristic elements include, above all, certain conceptions of time and space. Kant's aprioristic philosophy of time and space would be hardly conceivable without the scheme-content dualism.

Of course, the scheme-content dualism does not necessarily lead to an aprioristic philosophy. Rudolf Carnap, for example, replaced Kant's unique scheme of apriori concepts, allegedly representing invariant features of human understanding, by the multitude of possible language forms, which – allegedly – can be freely chosen on purely pragmatic grounds.[1] One may be led, then, to the view that important theoretical changes in the development of the sciences are distinguished by a change of the languages of the theories involved and that this change of languages is accompanied by a reorganization, or restructuring, of the material of experience.

In a recently published book about Wittgenstein the philosopher David Pears expresses this view as follows: "Scientific theories do not always invoke new things discovered at a deeper level [as, for example, in particle physics]. A newly discovered pattern in the behaviour of familiar things can be just as explanatory. This alternative was discussed by Wittgenstein in the *Tractatus*, and he used Newton's treatment of gravity to illustrate it, but perhaps the example that would occur to most people today is Einstein's special theory of relativity, which simplified our picture of familiar phenomena by redefining simultaneity" (Pears 1988, 203). What Pears apparently has in mind is that Newton's theory somehow succeeded in imposing upon the domain of familiar phenomena a unique, absolute relation of simultaneity, unrelativized to any inertial frame, whereas Einstein's theory, by a so-called 'redefinition', restructured the domain of familiar phenomena by means of a class of relativized relations of simultaneity.

This sounds like magic. As if we were able – by a ceremony called "definition" – to imprint on nature an absolute relation of simultaneity, and later on, when we are dissatisfied with this relation, to carry out some new imprinting on nature leading to the class of Einsteinian relativized relations.

The important issue that is touched upon here is the problem of how language relates to the world – or, more precisely, the problem of *reference*: how

---

[1] See (Carnap 1934), § 17.

words are related to objects, how they refer to objects. The term "object" should be taken here in its broadest sense, so that, for example, also physical events count as objects. Let us consider, for instance, the word "simultaneous". Its reference – or, speaking more technically, its extension – is a two-place relation, the relation of simultaneity, on the set of physical events. The question, then, arises: What is the extension of the predicate "simultaneous" in Newtonian physics? Is it well-determined? And what happens to the extension in the transition from Newtonian physics to special relativity theory?

Obviously, the extensions of our predicates are structuring a certain domain of objects (in the broadest sense of "object"). In the case of the predicate "simultaneous", for example, it is the domain of events, and then the view expressed by David Pears says, firstly, that this structuring is produced by certain definitions and, secondly, that scientific revolutions bring about a change in the extensions of our predicates, leading to a new structure on the domain of objects.

So far I have mentioned only views of philosophers. Let me now quote a physicist. In his paper "The Nature and Structure of Spacetime", Jürgen Ehlers says: "Einstein's examination of the verifiable, concrete meaning of the notion of time and his elimination of apparent contradictions from the electrodynamics and optics of moving bodies which resulted almost without effort from the substitution of an operationally meaningful time concept for a dogmatically postulated absolute time provided the model for a critical, empirically oriented re-examination of physical concepts in general"(Ehlers 1973). This is certainly true, and I actually don't know whether this statement is based on the content-scheme dualism. It leaves unclear, however, what we should say with respect to the reference of Newtonian concepts. What should we say, for example, about the predicate "simultaneous" as used by physicists in the eighteenth and nineteenth century? Consider their utterances which contain the word "simultaneous". Were they all empty? Did they refer to nothing real or specific?

I think I have quoted enough now. Perhaps we all tend to formulations of this sort and perhaps we all have at the backs of our minds this picture of some neutral empirical content waiting to be structured by a conceptual scheme. I now want to consider what this 'empirical content' (or this 'material of experience', or what have you) actually could be. I shall examine three possible answers. There is, firstly, the answer of traditional epistemology: the material of experience consists of sense data. Secondly, there is the answer of modern naturalized epistemology: the material of experience – that which supplies us with information about the world – consists of something neurological, for example, stimulations of our sensory receptors, or perhaps something in the interior of the brain. Thirdly, there is an answer which physicists – more precisely, relativists – may give: the material of experience consists of events. Events could be thought to be the theory-neutral and at the same time intersubjective stuff waiting to be structured by physical concepts (in particular by the concepts of space and time). Let us examine these three possibilities one after the other.

2. Traditional and Quinean Epistemology

The proposal of traditional epistemology – namely, that the material of experience consists of sense data – can be repudiated very quickly. There are numerous convincing arguments against the conception of 'sense data'. Let me simply remind you of some of them. First of all, we don't know what these objects, called "sense data", really are. Their identity and their way of existence are dubious to such an extent that they appear to be nothing more than a myth: the myth of the given, as Wilfrid Sellars has called it.[2] Furthermore, it remains unclear what sense data would be good for. The ambitious epistemological projects of reducing our normal talk about physical objects to talk about sense data all failed. Talk about physical objects cannot, even in the last analysis, be considered as highly derived talk about sense data (as, for example, Rudolf Carnap would have it in his book *Der logische Aufbau der Welt*). Even if sense data did exist, our normal talk about physical objects is uncoupled from them and has its own way of life beyond them. This is no wonder since sense data are essentially private. We have no use for them in our public language. Thus, the idea of sense data should be abandoned.

We can still try, however, to adhere to an epistemological orientation by lowering our ambitions. We can give up the quest for certainty and all attempts at phenomenological reductions, and we can try our hands instead at a naturalized epistemology. This was Quine's reaction to the failure of Carnap's reductionistic project.[3] Quine replaced Carnap's sense data, which are mental entities, by stimulations of our sensory receptors, which are something physiological. Sensory stimulations – or 'triggerings', as Quine likes to call them – are the empirical data of the modern materialistic epistemologist. They are his 'material of experience'.

Let us first ask: why stimulations of our sensory receptors, and why not something in the interior of the brain? Perhaps there is no big difference between these alternatives. The naturalized epistemologist is interested in a conception of 'empirical datum' which has to do with our contact with reality outside us. Through this contact we receive information about reality. Perhaps it doesn't matter very much whether we choose as contact places our sensory receptors or something causally more remote from these receptors inside the brain. Stimulations of sensory receptors recommend themselves by being relatively clear-cut happenings. Their choice as 'empirical data' doesn't appear to be too arbitrary, I think. So let us accept them as empirical data, as our material of experience.

---

[2] See (Sellars 1963), Essay 5.
[3] See (Quine 1969), Essay 3.

The domain of sensory stimulations, however, is not 'structured' by our concepts. The picture of *structuring* is not appropriate in this case. Quine uses, instead, the picture of *organization*. For example, with respect to the transition from Newtonian to relativistic physics, he says "that the new theory organized the pertinent data more simply than the old" (Quine 1990) (and the 'pertinent data' are, of course, certain sensory stimulations).

The domain of sensory stimulations cannot be said to be structured by our concepts since our concepts normally do not *refer* to sensory stimulations. Normally we talk about physical objects, and this talk is neither highly derived talk about sense data nor highly derived talk about sensory stimulations. Structuring has to do with reference to objects and not with stimulations of sensory receptors.

It is important, however, to understand the connection between reference and sensory stimulation. According to Quine, this connection has a purely instrumental character. In his important essay "Things and Their Place in Theories", Quine writes: "The scientifc system, ontology and all, is a conceptual bridge of our own making, linking sensory stimulation to sensory stimulation."(Quine 1981, 20) "Our talk of external things, our very notion of things, is just a conceptual apparatus that helps us foresee and control the triggering of our sensory receptors in the light of previous triggering of our sensory receptors. The triggering, first and last, is all we have to go on."(Quine 1981, 1) In other words: theories, and also the objects referred to by theories, are nothing but instruments that help us foresee sensory stimulations.[4]

One of the most remarkable consequences of this view is Quine's thesis of the *indeterminacy of reference* (or, to use Quine's more picturesque word, the *inscrutability of reference*).[5] If the objects referred to by a theory are seen only in their instrumental function with respect to predictions, their individuality becomes unimportant. Really important, then, is only the structure the theory imposes upon a domain of objects, while the objects themselves can be chosen arbitrarily.

To see this more clearly, let us consider the complete set of sentences which express what we believe about the world. Call this set "our total theory of the world". There are complicated logical relations between these sentences, and they allow inferences to observation sentences which give empirical significance to our theory.

What is an *observation sentence*? An observation sentence, in Quine's sense, is a sentence like "Here is a piece of chalk" which is directly correlated to certain characteristic sensory stimulations, such that we assent to it, or dissent from it, solely on the basis of these stimulations and quite indepen-

---

[4] Usually instrumentalists *deny* that theoretical terms refer (see, for example, (Vollmer 1991), 132). According to Quinean Instrumentalism theoretical terms are allowed to refer; but now reference itself is regarded as an instrument for prediction only.

[5] See, for example, (Quine 1981), 19–21.

dently of any considerations that go beyond the observational level.[6] Quine thinks that it is the sentence as a whole that is correlated to sensory stimulations, and not single words of the sentence. This sounds plausible since the reactions of assent and dissent caused by sensory stimulations are indeed directed towards sentences and not towards words. Assent or dissent concerns sentences and not words. Thus, the contact between language and sensory stimulations is, as Quine has put it, a 'holophrastic' one.

Let us come back to our total theory of the world. Let us ask what we can know about the reference of its descriptive terms and about its ontology, that is, about the domain of objects referred to by its descriptive terms. Suppose that philosopher A claims that a certain domain O_1 is the ontology of our theory. The descriptive terms, then, refer to objects in O_1 and through this reference the sentences of our theory receive their truth values (that is, they turn out to be true or false). Now philosopher B comes along and replaces each object of O_1 by a certain substitute and in this way produces a domain of objects O_2. O_2 may be equal to or different from O_1. Furthermore, philosopher B claims that the descriptive terms of our theory actually refer to his substitutes of the old objects and that (if O_2 is different from O_1) the true ontology of our theory actually is domain O_2. He can do this with good reason. His new reference relation obviously does not change the truth values of the sentences of our theory: truth values of sentences depend only on the structure which the extensions of the descriptive terms carve into the domain of objects, and philosopher B's substitutions of objects obviously do not change this structure. Of course, they also do not change the logical relations between the sentences. Furthermore, they do not change our theory's relation to the empirical data. This relation consists in the holophrastic contact of observation sentences to sensory stimulations, and it is completely independent of the reference of words. Consequently, the different reference relations of philosophers A and B are completely equivalent. There is no reason to prefer one to the other. Thus, reference appears to be indeterminate. This is what Quine calls "the inscrutability of reference".

Note that the indeterminacy coming to light through this argumentation cannot be diminished by observation or by ostension. Pointing to a piece of chalk while uttering the sentence "This is a piece of chalk" at best can correlate the sentence as a whole to certain sensory stimulations, but not the word "piece of chalk" to the object (namely, a piece of chalk). Believing the contrary means believing in magic. (This sort of magic, by the way, seems to have found its expression in the English word "spell". To spell the word "piece of chalk" while pointing to a piece of chalk: does this mean to put the spell on that piece of chalk such that the word becomes somehow connected to it?) Naming, referring to objects, is a ticklish affair. As Wittgenstein remarked in his *Philosophical Investigations*, the act of naming may all too easy be seen as a sort of occult process. Naming may appear, then, as some queer connexion

---

[6] See, for example, (Quine 1990), Chapter I and §§ 14–16.

of a word with an object (Wittgenstein 1953, § 38). Although Quine himself behaves a bit mysteriously when talking about the 'inscrutability' of reference, his true intention actually lies in clearing up all appearances of occultness or queerness.

Let us look at observation sentences – for example, the sentence "This is a piece of chalk" – a bit more closely. According to Quine, the meaning of an observation sentence, relative to a person who carries out the observation, is a pair of classes of sensory stimulations of that person: the class of stimulations that prompt assent to the sentence, and the class of stimulations that prompt dissent from the sentence.[7] The objects the sentence is about – pieces of chalk, for example – don't play any role in this conception. It's only sensory stimulations that matter.

Although sensory stimulations belong to single persons, Quinean observation sentences nevertheless fulfil the requirement of intersubjectivity: if person A assents, for example, to the sentence "This is a piece of chalk" in a certain perceptual situation, then, in the same perceptual situation, also person B assents to this sentence. According to Quine, this intersubjectivity is what makes science objective (Quine 1990). Quine's notion of objectivity is typical for an instrumentalistic position: objectivity is simply equated with intersubjectivity.

In the case of observation sentences, this intersubjectivity is accomplished by the process of language learning, which is a process of adaption. The child adapts itself to the linguistic behaviour of the speech community. Observation sentences can be learned in very early stages of this process, where the linguistic tools for reference to objects are not yet existent. Quine wants to understand how we acquire these tools, and that's one reason why he has devised a notion of meaning for observation sentences that does not involve reference to objects. Quine has a relatively detailed theory about these linguistic tools[8] – which I cannot go into here – and this theory, in Quine's opinion, shows what's the real substance of the notion of reference.

Beyond these linguistic tools, however, reference appears to be without substance in Quine's philosophy. From an epistemological point of view it appears, as we have seen, to be extremely indeterminate; and from an ontological point of view it appears to be trivial: the word "piece of chalk" refers to pieces of chalk; more cannot be said. This is the Quinean 'redundancy view' of reference. Quine is led to this depreciation of the concept of reference by his desire to avoid any magical view of reference. His defence against magic consists in accepting only a purely naturalistic, or even physicalistic, worldview. He thinks that within this worldview reference cannot be considered something substantial.

---

[7] Quine calls this pair of classes of sensory stimulations the *stimulus meaning* of the sentence in question; see (Quine 1960), §§ 8–10.

[8] See (Quine 1974) and (Quine 1984).

I must say that I find this depreciation of reference hardly acceptable. I think it worth the effort to look for questions about reference that are more promising than the Quinean ones.

3. Sameness of Reference

One question that hardly ever appears in Quine's philosophy concerns the change of our beliefs: what's happening with reference if theories change, in particular if they change in a revolutionary way? This is an important question in the philosophy of Thomas Kuhn, and also, for example, in the philosophy of Hilary Putnam. Putnam, for example, asks: what's happening with the reference of the term "electron" in the transition from Niels Bohr's early theory of the atom to quantum mechanics?[9] This is a question about the identity or constancy of reference: did the extension of the word "electron" change or did it remain the same?

Questions of this sort have been answered differently by different people. Thomas Kuhn very often sees a change of referenc; Hilary Putnam, on the other hand, very often sees constancy. I tend to the latter view, and in what follows I want to give several arguments for it. These arguments will throw light on my main subject: the dualism of content and scheme.

To begin with, one must say that Thomas Kuhn, when claiming changes of reference, very often seems to contradict scientific practice, which, among other things, is an interpretative practice. Scientists interpret utterances of former scientists from the viewpoint of their best theories, and in doing so they presuppose as far as possible that reference has not changed. If a physicist of our times looks at Niels Bohr's early theory of the atom he will presuppose, I think, that Bohr's term "electron" refers to the same entities as his own term "electron". (Whether these entities should be called "particles" is another question.) Of course, Bohr himself made this presupposition of sameness of reference, when using the same word "electron" in his early as well as in his later theoretical endeavours.

To be sure, Bohr's early statements were different from his later statements. This difference of the statements, however, can be explained without postulating a difference in the extensions of the words the statements are compounded of. One can see the actual importance of the so-called *causal theory of reference*, developed by Saul Kripke and Hilary Putnam[10], in its endeavour to devise explanations of exactly this sort. At the moment, I will not go into the details of this theory. I shall explain some of them later on. At the moment, it may be enough to say that the causal theory of reference should be

---

[9] See (Putnam 1975), 197, and (Putnam 1988), 12f.

[10] See (Kripke 1980) and (Putnam 1975), Essays 11–13. See also the improved version of this theory devised by Philip Kitcher, for example in (Kitcher 1978) and (Kitcher 1982).

seen as embedded in a general theory of *interpretation*, and that it is nothing else but a theory of the constancy of reference over time. The causal theory of reference doesn't want to say what reference *is*; rather, it is drawing a picture which shows how the interplay of human action and the human environment gives rise to constancy of reference. This interplay is a causal one, and that's why the theory is called the "causal theory of reference".[11]

What can be said about Quine's indeterminacy of reference in the light of the causal theory of reference? If I see it correctly, this indeterminacy is not affected by considerations concerning the constancy of reference. Quine's indeterminacy thesis only says that in the light of certain epistemological reflections – reflections which address the question of what our empirical data permit us to know about reference – the extension of, for example, the word "electron" appears to be indeterminate, no matter whether this word is used by the early Bohr or by an adherent of quantum physics. Quine's thesis, however, does not exclude the possibility of claiming with good reason that this extension did not *change* in the transition from the early Bohr to quantum mechanics. In other words, questions about the constancy of reference can be understood *modulo* referential indeterminacy.

We have already seen that even so-called ostensive definitions do not alter Quinean indeterminacy. Ostensive definitions do not 'constitute' reference; their function is a more modest one. With Wittgenstein, we could say: don't think what the function of ostension must be, but look at the language games we actually play with ostensions.[12]. We shall see, then, that these language games justify talking about the sameness of reference. If the person beside me says, with a pointing gesture, "This is a piece of chalk" then, among other things, he shows me that with the word "piece of chalk" he refers to the same, or not the same, objects than I do. Such is the way we use the expression "to refer to the same objects".

So, the causal theory of reference has no effect on Quinean indeterminacy of reference, and consequently it cannot cure those people who want to overcome the indeterminacy by means of magic. As we shall see shortly, however, it is a good cure against magical views of a different sort. There are (or have been) people, for example the adherents of so-called *protophysics*, who accept the empirical findings of general relativity theory and who, at the same time, believe that they can nevertheless stick to the view that physical space is euclidean, traversed by euclidean straight lines. These people are turning the reference of the word "straight line" into a mystery, for, in the light of general relativity it is not to be seen how this word could be forced to a sort of

---

[11] If seen in this general way, as embedded in a general theory of interpretation, the causal theory of reference seems to be defensible against much criticism that has been directed against it since its emergence. For example, the so-called 'introductory events' (Putnam 1975, 200) lose the outstanding and somewhat mythical role which they initially had in the theory (and which is justly criticized, for example, in (Moulines 1991)).

[12] Compare (Wittgenstein 1953), § 66

reference which ignores the curvature of space, i.e., which ignores the actual behaviour of light rays, clocks and measuring rods.

4. Events and Interpretative Practice

Let me come back, after this digression on reference, to the dualism of empirical content and conceptual scheme. It is time now to examine our third possible answer to the question of what this so-called empirical content might be. In the context of physics it suggests itself to consider not sense data or sensory stimulations but something more objective. Let us consider events, as to be found in relativity theory. The set of events might be said to constitute a theory-neutral reality waiting to be structured by theoretical concepts (and in particular by the concepts of time and space).[13]

In order to understand this view more clearly, I want to concentrate on the transition from Newtonian to relativistic physics (which I will call "the Newton-Einstein transition"), and I want to address the following two questions. First, can the set of events be said to be 'theory-neutral? In particular, are we justified in presupposing that it is the same set of events that underlies Newtonian as well as relativistic physics? Second, what's the effect of the Newton-Einstein transition on the reference, the extension, of our concepts, in particular of the concepts of time and space?

One can think of numerous motives for examining the concept of an event in physics. I shall be concerned with this concept only as used in classical (i.e., nonquantum) physics, and I am only interested in its supposed neutrality with respect to different theoretical structures. There are at least the following two points that cast doubt upon this neutrality. Firstly, the standard explications of the concept of an event refer to physical processes, which certainly are something structured and whose descriptions normally are theoretical to a considerable extent. Later on, however, the set of events is treated as a structureless, universal basis ready to be structured by different physical theories, which themselves are devised in order to explain the aforesaid physical processes. There certainly is a tension between these explications, on the one hand, and the subsequent use that is made of events, on the other. Secondly, the set of events consists not only of actual events, but also, and in an essential way, of possible ones that are not actualized. It would be bad

---

[13]This view is expressed with particular clarity in (Geroch 1979), where on p. 3 we read: "an event is to be regarded as a part of the world in which we live, not as a construct in some theory". See also (Misner et al. 1973), 225: "The primitive concept of an event [...] needs no refinement. The essential property here is identifiability, which is not dependent on the Lorentz metric structure of spacetime." Compare with this, however, Robert Wald's statement in (Wald 1984), 213: "the notion of an event makes physical sense only when manifold and metric structure are defined around it". Wald's view is in harmony with my subsequent reflections. See also (Mühlhölzer 1989), section 4.4.

metaphysics, however, to consider the concept of a 'non-actualized but nevertheless possible event' as a theory-neutral one. Thus, the idea of there being one, theory-neutral set of events, which is the basis of both Newtonian and relativistic physics, sounds rather dubious.

I think, however, that there is something correct in this idea. It is certainly wrong to consider the set of events as some theory-neutral stuff waiting to be structured by our concepts (like a cake waiting to be cut into pieces). It is correct, however, to consider the set of events as the same in Newtonian as well as in relativistic physics. To see this, we have to turn to our second question, which concerns the reference of our concepts.

I have been talking about the causal theory of reference that allows the justification of claims about the sameness of reference. This theory turns the tables with respect to the content-scheme dualism. Not the so-called content is taken as primary, but the best theoretical scheme that we have at our disposal. The causal theorist's picture is not that there is some neutral set of events waiting to be structured by our developing theories, but that from the standpoint of our present best theory we have to interpret the utterances of former theoreticians. We are called upon to tell *interpretative stories* about former utterances, and these stories will reveal that only very rarely does a genuine change of reference occur.

In general, interpretations of this sort are not concerned with sentences in the abstract, but with concrete utterances of sentences in specific situations. These utterances have to be interpreted.

Let us consider, as an example, the concept of simultaneity. What can we say, from the viewpoint of special relativity, about the reference of this concept as used by physicists adhering to Newtonian physics (whom I shall call "Newtonian physicists")? Let us try to tell plausible interpretative stories about their utterances containing the word "simultaneous".

Of course, I cannot deal with all possible utterances in all possible situations. I shall be concentrating on one characteristic situation which corresponds to a certain method of determining the simultaneity of events. It may be objected that no Newtonian physicist actually used, or even thought of, such a method, because he did not consider it necessary to do so. This may be true. But any Newtonian physicist, if presented with such a method, would have accepted it as a legitimate method of determining simultaneity; and that's enough for my present purpose.

The method is this (and I must apologize for the martial scenery): Consider two identical guns positioned at the midpoint between two points A and B of an inertial frame S. Suppose that cannon balls, simultaneously fired from the guns towards A and B, actually hit A and B. Then, these two hitting events are simultaneous. This is true, from the point of view of special relativity theory, relative to any inertial frame S, and it would be accepted also by a Newtonian physicist. The Newtonian physicist, however, believes that this method leads to the same relation of simultaneity for each inertial frame S, such that a relativization to S proves to be unnecessary. From the viewpoint of special relativity this is an erroneous belief. It can be explained quite easily, however, by taking into consideration that Newtonian physicists dealt with inertial frames whose relative velocities were only very small, and that they had no pressing theoretical reason to question their familiar spacetime framework. Thus, it seems perfectly plausible to say that in the situation of the cannon ball method Newtonian physicists, when using the word "simultaneous", refer to the relation of simultaneity relative to that inertial frame in which the guns are at rest. This interpretation, given from the point of view of special relativity, allows satisfactory explanations of the beliefs and the linguistic behaviour of Newtonian physicists.

Interpretations of this sort can be given for all situations relevant to the concept of simultaneity, and they can be given for any concept whatever. I think that the most plausible intepretations will be those which only very rarely postulate changes of reference.[14] This is particularly true, I think, of the concept of an event. Our interpretations will reveal that we are indeed justified in presupposing that the set of events is the same in Newtonian and relativistic physics.

An interpretative enterprise of this sort leads to the following view of Nature. Nature consists of those theoretically postulated objects and relations which survive theoretical changes. Or, to take another metaphor, Nature – or the world – 'crystallizes' in our theoretical development. What is crystallizing is sameness of reference of terms which belong to successive theories. And since the theories become better and better (at least in an instrumental sense) we are justified in saying that they lead to more adequate knowledge about the objects their terms refer to. This can be seen as a development towards more and more *ontological objectivity*. The concept of ontological objectivity in this sense is not an instrumentalistic one – like, for example, the Quinean concept of objectivity as intersubjectivity – nor is it a metaphysical concept. It simply concerns the constancy of reference in the development of our theories.

What is the role of observation and measurement in this view? As our simultaneity example shows, methods of empirical determination give important clues to interpretation. In contrast to an operationalist view, empirical determinations are now seen not as means of defining our concepts, but as means of interpreting them, and the interpretative enterprise allows us to assert the constancy of reference in spite of changes in the methods of empirical determination.

What, then, of the so-called 'world of Newtionan physics'? In the course of the interpretative enterprise based on relativistic physics, this world seems now to have disappeared. It has disappeared, in fact, as a 'world', because there is only one world – at least only one world of physics – and this one world is not a Newtonian one. The so-called 'world of Newtonian physics' has not disappeared, however, as a *model* of Newtonian physics. Models of theories are

---

[14]The interpretative stories which I have given in (Mühlhölzer 1989), Sect. 7.4, confirm this assessment.

truth-makers: they make the axioms of the theories true. However, reference, which has been my main subject, has a different status. It gives rise to truth and falsity. In other words, relations of reference are only truth-value makers. We all know that our actual theories, given the actual reference of our terms, are not completely true. Thus, reference to objects must not be confused with delivering models.

5. Why Sameness of Reference?

I now want to look more closely at the interpretative practice which makes us see constancy of reference. A convinced instrumentalist may be unimpressed by this practice. He may ask, what's the significance of it? Why do we postulate constancy of reference? Niels Bohr's early theory of the atom, containing the term "electron", as well as quantum physics, containing this same term, possess certain empirical consequences, and measured against these consequences quantum physics turns out to be preferable. Why, then, ask whether the term "electron" occurring in these two theories refers to the same entities? Is it not quite unimportant whether we say "yes" or "no" to this question?

Let us look at the scientist in a Quinean way. According to the instrumentalist Quine, Niels Bohr and the physicist of today are nothing but information processing systems – or, more precisely, stimulation processing systems – which foresee future sensory stimulations in the light of previous ones. Sentences containing the word "electron" help them to do this in an effective way. It doesn't matter, however, whether this word, as used by Niels Bohr or as used by the physicist of today, refers to the same or not the same objects. What matters is whether it helps us foresee sensory stimulations, and this help, as we have seen above, does not depend on the objects themselves but only on structural issues. The Quinean subject does not have contact with things, but with sensory stimulations, and it is no wonder, then, that in Quine's philosophy reference to things appears to be indeterminate and that questions about the constancy of reference don't play any role.

Obviously this Quinean view is at odds with our normal way of looking at and explaining human action, and in particular linguistic action. Our normal explanations of action refer to the beliefs and intentions of persons and take into account their dealing with things – and not with sensory stimulations. Explanations of this sort are an essential part of the interpretations I have been talking about. If we want to interpret the utterances of Niels Bohr, we have to take into consideration Bohr's beliefs and intentions and Bohr's dealings with things in his environment.

What exactly is the role of things in explanations of this sort? It's essentially a *causal* role. The presence of an object in the perceptual range of a person is causally relevant to the formation of beliefs and the production of linguistic utterances of this person. On the causal path from the object to the beliefs and utterances, however, we also find the Quinean sensory stimulations.

They are an intermediate stage in this path and they also are causally relevant to beliefs and utterances. The question arises of whether Quine's physicalistic point of view, dealing with triggerings of sensory receptors, can be reconciled with the point of view of our normal explanations of action, which are interpretative enterprises dealing with things and which, at least prima facie, seem to be far away from a physicalistic worldview.

This reconciliation will only be possible if both views make use of the same conception of causality. Whether this is the case, I don't know. Maybe in the distant future the causal role of sensory stimulations in the production of linguistic utterances will be understood within a physicalistic framework. It could be, however, that the causal role of the objects themselves, which are referred to by our utterances, can only be understood within an interpretative enterprise that goes beyond any physicalistic framework.

If this were the case – as has been argued, for example, by Hilary Putnam[15] – then we should become accustomed to the idea that a causal account of reference forces us to abandon our scientific worldview, and my own considerations, which tried to rescue the unity of this worldview, were situated outside its border. Note that this result could be seen as grist for Quine's mill. Of course, Quine would look at it the other way round: instead of abandoning the scientific worldview, he prefers to abandon the idea that reference has to do with causality. From the very beginning, Quine would have had a good nose for the sacrifices to be made on the altar of scientism.

It seems to me, however, that these sacrifices are a bit too extensive. Scientism à la Quine seems to sacrifice the scientist himself. Let us ask again: Why should we postulate constancy of reference? My answer so far has been that these postulations belong to our normal *interpretative practice*. But what's the point of this practice?

One important point is, I presume, that the factual development of our theoretical constructions depends on this practice. I think that theoretical innovations are essentially based on the theories already at our disposal. Not only do we interpret the former theories in the light of the later ones (as I stressed before); at the same time we also need the former theories in order to know what we are actually doing in the later ones. In particular, we need the objects referred to by former theories as *fixpoints of our research*. This subtle and important – and, admittedly, not very well understood – interplay between former and later theories[16] is totally ignored in the Quinean instrumentalistic picture. According to this picture our theoretical development is nothing more than a process of adaption: theoretical constructions have to be adapted to empirical data, which are seen as the *only* fixpoints of our research. Intertheoretical relations appear to be unimportant. This picture, however, which apparently is suggested by the third dogma of empiricism,

---

[15] See (Putnam 1988), Chapter 7.

[16] I shall deal with this interplay, which was also the main topic of (Mühlhölzer 1989), in a forthcoming book.

is an empiricist fiction. If we try to leave fictions and turn to real life, intertheoretical relations become important and with them questions about the sameness of reference. Then the notion of reference gains substance.

6. Conclusion

Let me come back, finally, to my main subject, the dogma of an empirical content waiting to be structured by a conceptual scheme. The foregoing considerations suggest, in accordance with the result of Donald Davidson, that this dogma should be rejected. It is particularly misleading with respect to the development of our theories, where it should be replaced by a view based on the idea of *sameness of reference*. According to this view, the fixpoints of our research are not some theory-neutral empirical data, to which our theories have to be adapted, but the objects referred to by our theories. The picture of content and scheme is not needed. This picture should be rejected even in that case which seemed to be the most appropriate for its application, namely the case of the concepts of time and space. The concepts of time and space certainly possess a privileged position within the conceptual framework of physics. This position, however, does not come from a mysterious faculty of structuring some so-called material of experience.

Acknowledgement

This paper owes much to discussions following the presentation, at the 4th Philosophy and Physics Workshop at FEST, Heidelberg. I should like to acknowledge the useful critical comments of the participants of the workshop. Furthermore, I am grateful to Marshall Kavesh for correcting and improving my English.

References

Carnap, Rudolf (1961), Der logische Aufbau der Welt (1928), Hamburg: Meiner

Carnap, Rudolf (1934), Logische Syntax der Sprache, Wien: Julius Springer

Davidson, Donald (1984), Inquiries into Truth and Interpretation, Oxford: Clarendon Press

Ehlers, Jürgen (1973): The nature and structure of spacetime. In: J. Mehra (ed.), The Physicist's Conception of Nature, Dordrecht: Reidel

Geroch, Robert (1978): General Relativity from A to B, Chicago: University of Chicago Press

Kitcher, Philip (1978): Theories, theorists and theoretical change. In: Phil. Rev., 87, 519–547

Kitcher, Philip (1982): "Genes", in: Brit. J. Phil. Sci., 33, 337–359

Kripke, Saul (1980): Naming and Necessity, Cambridge, Mass.: Harvard University Press

Misner, Charles W., Thorne, Kip S., Wheeler, John A. (1973): Gravitation, San Francisco: W. H. Freeman

Moulines, C. Ulises (1991): Über die semantische Explikation und Begründung des wissenschaftlichen Realismus. In: H. J. Sandkühler/ D. Pätzold (eds.), Die Wirklichkeit der Wissenschaft – Probleme des Realismus (Dialektik 1991/1), Hamburg: Meiner, 163– 178.

Mülhölzer, Felix (1989): Objektivität und Erkenntnisfortschritt, Habilitationsschrift, München

Pears, David (1988): The False Prison: A Study in the Development of Wittgenstein's Philosophy, Vol. II, Oxford: Clarendon Press

Putnam, Hilary (1975): Mind, Language and Reality, Phil. Papers, Vol. 2, Cambridge: University Press

Putnam, Hilary (1988): Representation and Reality, Cambridge, Mass.: MIT Press

Quine, Willard Van Orman (1960): Word and Object, Cambridge, Mass.: MIT Press

Quine, Willard Van Orman (1961): From a Logical Point of View, Second Edition, New York: Harper and Row

Quine, Willard Van Orman (1969): Ontological Relativity and Other Essays, New York: Columbia University Press

Quine, Willard Van Orman (1974): The Roots of Reference, LaSalle, Ill.: Open Court

Quine, Willard Van Orman (1981): Theories and Things, Cambridge, Mass.: Harvard University Press

Quine, Willard Van Orman (1984): Sticks and Stones; or, the Ins and Outs of Existence. In: L. S. Rouner (ed.), On Nature, Notre Dame: University of Notre Dame Press, 13–26

Quine, Willard Van Orman (1990): Pursuit of Truth, Cambridge, Mass.: Harvard University Press

Sellars, Wilfrid (1963): Science, Perception and Reality, London: Routledge & Kegan Paul

Vollmer, Gerhard (1991): Wider den Instrumentalismus. In: A. Bohnen, A. Musgrave (eds.), Wege der Vernunft, Tübingen: J.C.B. Mohr (Paul Siebeck), 130–148

Wald, Robert (1984): General Relativity, Chicago: University of Chicago Press

Wittgenstein, Ludwig (1953): Philosophical Investigations, Oxford: Blackwell

On the Mathematical Overdetermination of Physics

Erhard Scheibe

Am Moorbirkenkamp 2a, D-22391 Hamburg, Germany

I

In the title of this paper I speak of the "mathematical overdetermination of physics", and the reader is fully entitled not to know what I mean by this. Right at the beginning of this paper it may be of some help to compare its subject with the related though different so-called theoretical overdetermination of a corpus of empirical data. Just as an empirical theory often exhibits an unnecessarily rich structure when compared with the observational data to be explained by it so the mathematics introduced to formulate a physical theory frequently brings a wealth of structure into play that cannot be matched by the physical elements of that theory. In both cases we find ourselves deluded in our expectation that in order to reformulate a certain corpus of statements by submitting it to logical analysis there be only two things to be taken into consideration: 1) the concepts characteristic for the corpus in question, and 2) the logical expressions binding together those concepts. I say we are deluded in expecting this because in both cases of overdetermination the truth seems to be that a *third* component has to be considered. In the case to be dealt with in the following this phenomenon has found its expression already long ago in Galileo's saying that "the book of nature is written in the language of *mathematics*." Let me illustrate the situation by two examples: empirical laws and electrodynamics.

An empirical law establishes a relation between physical quantities by giving a numerical relation isomorphic to the former. By a gas law, for instance, the physicist wants to express a relation between pressure, volume and temperature of a gas. Though united in one and the same gas the quantities in question are of entirely different kinds, and no further physical concepts are available in order to bridge the gulf. In this situation we take a loan from mathematics. There is one thing that our quantities have in common: Their values can be described – one by one – by real numbers. With one stroke this uniformisation makes possible what seemed impossible before: the wealth of ternary relations between numbers is at our disposal to formulate the law. However, for this gain we have to pay a price. Whereas the mathematical relation chosen can be understood by its definition in terms of the elementary arithmetic operations no corresponding physical understanding is possible. It is not that we do not understand what is meant by the law in order to test it. But the arithmetic operations defining the law having received no physical interpretation, the truth conditions of the physical relation assumed by the law cannot be traced back to elementary *physical* facts as is possible for the corresponding mathematical relation with respect to elementary mathematical facts. Rather our physical law bears the burden of a piece of seemingly uneliminable surplus mathematics.

For my second example I refer to the well known fact that there are two approaches to electrodynamics: 1) via the potentials and 2) via the field strengths. Their equivalence is expressed by the theorem that the field strengths satisfy Maxwell's equations if there are potentials satisfying *their* equations and related to the field strengths in the usual way. Now, either approach, taken by itself, is heavily infected by mathematics, and there are no *a priori* reasons why one of them should be more physical than the other. It is only by experience that we have learned that two potentials leading to the same field strengths seem to be physically indistinguishable. Consequently, the field version of electrodynamics is to be preferred for physical reasons, and the version in terms of potentials can be viewed as a mere mathematical description of greater convenience in certain respects but introducing a structure much too strong to be justified on purely physical grounds.

The introduction of the electrodynamic field *instead* of the potentials can be viewed as the solution of the problem of eliminating an unnecessarily strong element of mathematics from a physical theory. The essential step in this elimination would be the complete detachment of the field strengths from the potentials in finding their own dynamics in the form of Maxwell's equations. By contrast, the case of the gas law stands unsolved, and it appears to be almost a miracle that the physical behavior of a dead piece of matter can be described by such a simple mathematical formula. The Pythagoreans, discovering the isomorphism between musical intervals and simple numerical relations, may have been the first to have experienced such a miracle. Since then the issue has repeatedly puzzled many great scientists. In our times Wigner said "that the unreasonable effectiveness of mathematics in the natural sciences is something bordering on the mysterious"[1]. Einstein speaks of the "enigma that has worried researchers of all times so much: How is it possible that mathematics, a product of the human mind independent of any experience, so excellently fits the objects of physical reality?" [2]. And for Steven Weinberg "it is positively

---

[1] Wigner (1979), pp. 223 and 229f

[2] Einstein (1934), pp. 119f

spooky how the physicist finds the mathematician has been there before him or her." [3].

Alongside such awestruck utterances there are also statements that would rather downgrade the phenomenon and view it in a more sober attitude. Such is the case with P. W. Bridgman in his *The Nature of Physical Theory* [4]. In this book we find an explicit discussion of exactly the point that I called the overdetermination of physics by mathematics. Bridgman is particularly impressed by the extension that this overdetermination has assumed in quantum theory. With respect to it he says:

The mathematical structure ... has an infinitely greater complexity than the physical structure with which it deals. In our elementary and classical theories we have become used to discarding perhaps one-half of the results of the mathematics ... But here we retain only an infinitesimal part of the mathematical results, and except for a few isolated singular points relegate the entire mathematical structure to a ghostly domain with no physical relevance.

On the other side Bridgman is not willing to become puzzled about this situation. He is aware of the tradition that I just alluded to:

The feeling that all the steps in a mathematical theory must have their counterpart in the physical system is the outgrowth ... of a certain mystical feeling about the mathematical construction of the physical world. Some sort of an idea like this has been flitting about in the background ... of the thinking of civilization at least since the days of Pythagoras, and every now and then ... it bursts forth again ... as in the recent fervid exclamation of Jeans that "God is a mathematician".

However, in Bridgman's view,

There would seem to be no necessity ... that all mathematical operations should correspond to recognizable processes in the physical system .... All that is required of the theory is that it should provide the tools for calculating the behaviour of the physical system, and it is capable of doing this if there is correspondence between those aspects of the physical system which it engages to reproduce and *some* of the results of the mathematical manipulations.

Behind Bridgman's considerations a hidden nominalism is at work. But as a physicist Bridgman is simply not interested in the question whether or not the mathematics actually used in physical theories is really necessary. The question of eliminability has recently been attacked most forcefully by Hartry Field. In his book *Science Without Numbers* Field proposes to show "that the mathematics needed for application to the physical world does not include anything which ... contains references to ... abstract entities like numbers, functions, or sets" [5]. Explaining his position in more detail Field says: "Very little of ordinary mathematics consists merely of the systematic deduction from axiom systems: My claim however is that ordinary mathematics can

---

[3] Weinberg (1986), pp. 725 and 727
[4] Bridgman (1936) pp. 116f, 67, 66 and 65
[5] Field (1980), pp. 1f

be replaced in application by a new mathematics which does consist only of this" [6]. Substantial work to the same effect of eliminating some mathematics from physical theory has also been done by Günther Ludwig (Ludwig 1990), however, he himself does not claim and presumably would not even agree to have done just this. It is here where the distinction between physical and mathematical terms borders on the different, though related and well known dichotomy between the so-called empirical or observational terms and the purely theoretical ones. However, in the following I do not want to clarify the relation of these distinctions any more than by what is already achieved by a clarification of the first distinction alone. The task before us, then, is to answer the question: What is the distinction between mathematical and physical terms (or entities) in a physical theory, and to what extent have we to accept the occurrence of the former in a theory that is called "physical" after all?(Cf. Scheibe 1986, 1992)

II

Setting about to answer this question, perhaps the best method is to take the bull by the horns and accept the view that in a physical theory we usually talk about mathematical entities just as well as about physical ones. It will be wise, however, to pay at least some attention also to the opposite view. This view is that there simply are no mathematical entities by themselves or, if there are and if we even could profitably introduce them into physical theory, we need not do so. An extreme version of this position holds that mathematics is merely a method – the method, for instance, of axiomatizing a certain body of knowledge. This view would be realized if we could produce a reformulation of whatever physical theory we know of within *first order logic* with all nonlogical symbols being interpreted directly by physical entities. A physical theory thus reformulated would not even contain genuine mathematical statements – let alone be about mathematical entities.

For the time being I take this extreme view to be unrealizable. In spite of the admirable efforts made by Field the view is unrealizable, certainly for technical reasons and probably for reasons of principle as well. Present theoretical physics is formulated by making reckless use of modern mathematics, and although, as we shall see, some purifications are possible, the total elimination of the mathematics embodied in physics would put us far beyond our present capabilities – not to mention the question whether a total elimination program is desirable after all.

Mathematical entities would still be avoided if we succeeded in reformulating physics by taking refuge with *higher order logics*, the non-logical symbols still being directly interpreted by physical entities alone. According to Gödel's incompleteness theorem there is then ample opportunity to make quasi-logical

---

[6] Field (1980), p. 107, n. 1

assumptions, e.g., comprehension axioms, alongside the physical axioms as far as necessary to develop the theory. Mathematics could then be identified with these quasi-logical axioms, and there would thus be genuine mathematical statements about non-mathematical entities. I think there is a chance for such an undertaking to lead to some successes. We could deal with physical structures of higher order, as they are at least very convenient in the more advanced theories of physics, and we could make higher order statements on first order structures, as they are needed in all cases involving a continuum.

With our next step we still remain within the bounds of first or higher order logics but now allow for the explicit introduction of *mathematical entities* into the theory. In a many-sorted version of the logic chosen this would mean assuming that one or the other sort of variables run over the natural numbers or the real numbers or a real number space of any dimension or the complex numbers or what have you. If the remaining sorts of variables are given a physical interpretation we already have a very powerful instrument at our disposal, and I am pretty sure that as far as physics can be axiomatized at all it could be axiomatized within this frame. However, once we have assumed higher order logics *and* non-physical descriptive terms we have, so to speak, crossed the Rubicon, and the question becomes urgent why, having gone that far, we do not go also the last step and introduce set theory right at the beginning.

There are at least four reasons to take advantage of this opportunity. In first and higher order logics the languages are usually interpreted by abstract structures, i.e., by certain systems of sets. Sets, therefore, come in anyway, and it is suggestive to make them the object of investigation quite explicitly. We would thus obtain with one stroke what otherwise would be their piecewise introduction, structure by structure, as the nature of the case demands it. As regards logic we would not have to go beyond the first order, and yet the whole stock of mathematics would be at our disposal. So if we want to analyze the role of mathematics in physics a reconstruction on the basis of set theory seems to have considerable advantage even if our final goal is the controlled destruction of this edifice by gradual elimination of its mathematics.

Now *set theory* as a rational reconstruction of mathematics is a common place. But what about physics? How would the description of a hydrogen atom as a *physical* system look like in the framework of a set theory? What sets do we meet with in physics? Let recall us first that our common understanding of what sets are is entirely neutral with respect to the distinction between abstract and concrete entities. There are sets of cows in exactly the same sense as there are sets of numbers, and I am a member of the assemblage of people now in this room in exactly the same sense as 3 is a member of the set of prime numbers. So sets and membership are universal but the sets we meet with in daily life more often than not are sets of concrete entities. We are even familiar with sets mixed from both abstract and concrete entities as, for instance, a list enumerating the individuals attending this meeting would be. And we are certainly less familiar with lists of abstract entities. From the viewpoint of daily life set theory seems farther away from mathematics than from any science concerned with bodily beings and matters of fact.

At this point, however, it has to be emphasized that the sets we meet with in physical theory usually are not sets of real things. They are sets of real possibilities, for instance, the set of events considered in special relativity or the set of states of a quantum mechanical system. Although we commonly speak simply of events and states, what we mean are possible events and possible states. A physical theory, even in its application to an experiment actually performed in the laboratory, is universal precisely because it transcends reality. It gives us an explanation of the experiment by telling us how this particular real system *might* have behaved had the determining conditions been other than in fact they were. It is because of this kind of use made of sets that a set-theoretical reconstruction of physics is not trivial and becomes related to mathematics.

To put in some technical detail, we may recall that a set theory explicitly introducing non-mathematical entities was developed by Zermelo already in 1908.[7] Zermelo provided for entities in a set- theoretical universe different from the empty set but like it in having no elements. These individuals, or *Urelemente* as Zermelo called them, are not needed in the construction of the universe of mathematical entities. They are free to be identified, for instance, with points of physical space or spacetime, with mass points, field strengths or other fundamental physical entities. Now in the usual Zermelo-Fränkel set theory containing an axiom of foundation there are no infinite chains

$$x_0 \ni x_1 \ni \cdots \ni x_n \ni \cdots . \tag{1}$$

This gives us the following tripartition of our universe of discourse. Given x_0 either all descending chains (1) cease with the empty set or all cease with an individual or some do the first and some the second. Accordingly, if all individuals are assumed to be physical, in the first case x_0 is mathematical, in the second physical, and in the third it is mixed. It can be shown that the physical sets can be constructed from individuals in a similar way as the mathematical sets are constructed from the empty set by means of the rank function. However, the most interesting sets in physics seem to be the mixed ones.

Our division of entities leads to a slightly more complicated situation for formulas. First of all, set-theoretical sentences, now being also about physical entities, assume an empirical status at least in principle. In this paper I shall not enter the question what it could mean that, for instance, the powerset axiom or the axiom of choice can be empirically refuted. The formulas of a particular physical theory are essentially *not* set-theoretical sentences but *open* formulas containing variables and additional constants indicating or referring to special and, most importantly, physical entities. A consistent system of open formulas can be made the axiom system of a conservative extension of

---

[7] For a modern presentation see Suppes (1960)

set theory. Recently, so-called species of structures have profitably been used to axiomatize physical theories (Cf. Ludwig 1990, Ch. 4; Bourbaki 1968). A species of structures in the sense of Bourbaki is a set-theoretical formula satisfying two essential conditions. Before giving them I should emphasize that the concept of a physical theory deserves much more attention than is paid by discussing the formal axiomatics. But for our purpose the restriction to this part of a theory is justified.

The first condition on a formula to be a species of structures is, as the name puts it, that the system of sets indicated by the free variables is a *structure* in the sense of mathematical logic. This means that the variables are divided into those indicating the base sets and those indicating the typified sets. The *typification* is expressed by the first conjunctive member of our formula, which has to be of the form

$$s_i \in \sigma_i(X_1, \ldots, X_m) \tag{2a}$$

where X_k are the base sets, s_i the typified sets and σ_i is a type, i.e., a prescription for how the set $\sigma_i(X_1, \ldots, X_m)$ is constructed from the base sets X_k by successively forming power sets of cartesian products. In this manner type-logical predication is simulated in set theory to deal with many-sorted structures of arbitrary types and (finite) orders. As regards the previously introduced division of all entities into physical, mathematical and mixed, it is reasonable to require that the base sets be either physical or mathematical. The typified sets then are generally mixed. The topology s of a topological space X, for instance, has the typification

$$s \in \mathrm{Pow}^2(X)$$

if s is taken to be the set of open subsets of X. Whereas s would be a physical set if X were one, the typification of the distance in a metric space X, as usually understood, is

$$s \in \mathrm{Pow}(X^2 \times \mathbb{R})$$

(with $\mathbb{R}$ as the set of real numbers) which makes s a mixed set. If, finally, we wish to express the triangle inequality for s we have to introduce addition into $\mathbb{R}$, and this gives us a mathematical entity typified by

$$ad \in \mathrm{Pow}(\mathbb{R}^3).$$

The second condition to be considered concerns the other conjunctive member of a species of structures: the *axiom proper*

$$\alpha(X_1, \ldots, X_m; s_1, \ldots, s_n). \tag{2b}$$

The condition is that α be invariant under arbitrary isomorphisms of the structure $\langle X; s \rangle$ – the mathematical sets being kept fixed –

$$\alpha(X; s) \leftrightarrow \alpha(X'; s'). \tag{3}$$

This invariance is automatically fulfilled by the typification, so our requirement can be paraphrased by saying that what is said by a species of structures is true or false of a structure no matter what the nature of the elements of the base sets of that structure is. This feature of a statement is sometimes taken to be the hallmark of its mathematical character. Whoever thinks about this seeks to decide whether we are dealing with something mathematical or something concrete at the level of statements as opposed to their subject. And indeed, since in the threefold universe introduced here every structure is isomorphic to a mathematical one we could, from a purely structural point of view, dispense with physical entities as explicit objects of our investigation and satisfy ourselves with mathematical descriptions of them. I am not of one mind whether this or the position taken here is the more adequate. There seem to be advantages and disadvantages on either side. In the following I stick to the two-sorted approach, which then calls for some caution in the manipulation of formulas because the characters we distinguish are obviously not invariant under isomorphisms.

III

The main problem that poses itself if physical theories are based on a set theory as their common logico-mathematical background is the problem of eliminating this basis in favour of one of the weaker frameworks that were mentioned earlier. This leads to deeper foundational problems that I shall not pursue in this paper. The great elimination problem presupposes intra-set-theoretical eliminations concerning the physical axiom systems, and it is this business that I want to duscuss in the last part of the paper. The *intra-set-theoretical eliminations* are re-axiomatizations including changes of the typifications of the respective species of structures. I will skip the definition of the notion of equivalence of two species of structures on which those changes are grounded, as it is straightforward. There are then two main kinds of eliminations of mathematical terms by equivalence transformations, one concerning defined constants, the other one bound variables occurring in the axioms.

Let me begin with a childishly simple example of the *first kind*. Suppose we have to deal with some population (X) containing two sorts of physical objects called "boys" (B) and "girls" (G). They may interact with each other by becoming friends (F). The typification of such a system would then be

$$B, G \subseteq X, \qquad F \subseteq X^2 \tag{4a}$$

and so far our axioms obviously are about physical beings and nothing else. But the substantial description of the interaction is still to come and will now be given by mentioning some cardinal numbers. We assume that each girl has two boyfriends, each boy has three girlfriends, and any two boys have one girlfriend in common. For the moment it suffices to pay attention to the fact that in these axioms proper of our theory *numbers* and therefore mathematical entities have been related to the physical existences assumed to be there. If anyone asked what our axioms are about, at least some people might find it quite natural to answer: They are about numbers *pari passu* with boys and girls. Accordingly, a set-theoretical formulation of, say, our third assumption would be

$$\bigwedge xy \in X. \quad x \neq y \rightarrow \widetilde{G_x \cap G_y} = 1 \tag{4b}$$

where generally $\tilde{m}$ is the cardinal of m and

$$G_x := \{y \in G \mid \langle x, y \rangle \in F \wedge x \in B\} \tag{4c}$$

Evidently, in this formulation the number 1 is mentioned just as the physical sets B, G, F and X are. Somebody subscribing to such a formulation must be prepared to answer the question: What is the referent of the symbol "1" in (4b)? However, in the case before us there is a way out for all who are scared by such a question. Our axiom system is equivalent to one in which no number terms occur in descriptive position. The number-free equivalent of (4b), for instance, would be

$$\left.\begin{aligned}
&\bigwedge xy \in X. \quad x \neq y \wedge x, y \in B \dotrightarrow \\
&\bigvee u \in X. \quad u \in G \wedge \langle x, u \rangle, \langle y, u \rangle \in F \wedge \\
&\bigwedge v \in X. \quad v \in G \wedge \langle x, v \rangle, \langle y, v \rangle \in F \rightarrow v = u \dots
\end{aligned}\right\} \tag{4d}$$

where only the physical sets are mentioned. At the same time you see that, for our little theory, this equivalence transformation is the first step in getting rid of set theory and replacing it by first order logic directly applied to the little scenario. As I said before, I do not want to embark on the great elimination problem – I might add that this is true not only in the case under discussion. The eventual difficulties and complications inherent in the basic background apparatus are now, after the transformation, to be found in the physical theory itself. One consequence, for instance, of our assumptions is that there are exactly four boys and exactly six girls. This can quite easily be proved within the set theoretical framework. By contrast, I know of no proof within our second, purely logical framework, and I am pretty sure that, though quite elementary, it would be fairly complicated.

My second example concerns euclidean geometry. Let us think of space and the point relations of congruence and betweenness as physical sets typified according to

$$CO \subseteq X^4, \qquad be \subseteq X^3 \tag{5a}$$

(X being the space). There may be a controversy about the precise meaning of saying that space points are physical entities. But in view of what is now to come we certainly can put aside all quarrels in this respect. In our axiom proper we introduce the real number space $\mathbb{R}$ by requiring that there exists a coordinate system φ on X such that the congruence and betweenness relations are carried into certain numerical relations according to

$$\langle xyuv \rangle \in CO \leftrightarrow \sum_i (y_i^\varphi - x_i^\varphi)^2 = \sum_i (v_i^\varphi - u_i^\varphi)^2 \qquad (5b)$$

and

$$\langle xyz \rangle \in be \leftrightarrow \left(\sum_i (y_i^\varphi - x_i^\varphi)^2 \right)^{\frac{1}{2}} + \left(\sum_i (z_i^\varphi - y_i^\varphi)^2 \right)^{\frac{1}{2}} = \left(\sum_i (z_i^\varphi - x_i^\varphi)^2 \right)^{\frac{1}{2}}$$
$$(5c)$$

respectively (x_i^φ being the components of x in φ). Again the details do not matter. All that is required from us now is to be impressed by the way in which formulas $(5b, c)$ give an answer to the question of what congruence and betweenness are like in euclidean space. In the spirit of analytic geometry the answer is given, not in physical, but in mathematical terms of numbers and algebraic operations on them. The mathematical overdetermination is evident from the group of euclidean transformations relating any two coordinate systems in which congruence and betweenness have representations according to $(5b)$ and $(5c)$ respectively. Yet the case of euclidean geometry is also a "solvable case". The tradition of synthetic geometry, going back to Euclid and culminating in the work of Hilbert and Tarski, has provided us with axiom systems equivalent to the foregoing one in which no non- geometrical entities are mentioned (Cf. Tarski 1959). As a representative example I mention the axiom of segment construction

$$\bigwedge xyuv \in X. \quad \bigvee z \in X. \quad \langle xyz \rangle \in be \wedge \langle yzuv \rangle \in CO. \qquad (5d)$$

in which both relations appear.

The solvability of the two foregoing examples does not mean that the subclass under discussion – mathematical sets mentioned in the axiom proper – is unproblematic in general. There are amazingly far reaching results indeed, for instance, the solution of Hilbert's fifth problem: The species of topological groups that are Lie groups is equivalent to the species of topological groups that are locally compact and have a neighbourhood of the identity containing no non-trivial invariant subgroup (Yamabe 1953). But in general the situation stands ill. Many physical theories, including classical and quantum mechanics, not only are extensions of euclidean geometry but usually are presented and developed by the method of analytical geometry. What has to be said on the essential physical entities is said, not in unquestionable physical terms, but in terms of coordinate representations. Numerical functions and systems of differential equations dominate the scene. The physical and mathematical elements of the theory are not clearly separated.

The *second subclass* of the kind of elimination problem under discussion has already been exemplified in the introduction. It is the case where mathematical sets are mentioned, not in the axiom proper, but in the typification of a theory. This means that the very object of the theory is a mixed structure. Such a sphinx is most typically brought about by making sets of coordinate systems part of the structure under investigation. Cases in point are canonical coordinate systems in Hamiltonian mechanics and spacetime coordinate systems in general relativity. If coordinate systems are meant for scaling physical quantities they are as a rule simply suppressed. But as the values of the quantities are then just real numbers, the set $\mathbb{R}$ of real numbers has to be made part of the physical structure. The apparently innocent introduction of coordinate systems into physical theory is in fact a hard nut to crack when seen from the viewpoint of mathematical overdetermination.

It is interesting to observe that the development of differential geometry has undergone a turn from coordinate representations of geometrical objects to so-called intrinsic or coordinate-free formulations. As far as this happened under the influence of physical applications the move to intrinsic representations seemed a move back to physical meaning. However, as long as the notion of a manifold based on local coordinates lurks behind the scene, the result of the movement is not completely convincing. The typification of, for instance, a linear connection ∇ in coordinate representation is given by

$$\nabla \subseteq \text{Pow}(M \times \mathbb{R}^n) \times \text{Pow}(M \times \mathbb{R}^{3^n}) \tag{6a}$$

where M is the (n-dimensional) manifold. If, on the other hand, the connection is viewed intrinsically as differentiation of a vector field by another vector field its typification becomes

$$\nabla \subseteq \{\text{Pow}\,[M \times \text{Pow}\,(\text{Pow}\,(M \times \mathbb{R}) \times \mathbb{R})]\}^3 \tag{6b}$$

which clearly shows that the situation, though improved in the relevant respect, is not completely restored.

In conclusion let me briefly look at the *other kind* of elimination problems. It concerns the bound variables in the axiom proper of a species of structures. In set theory quantifications are by definition over the whole set universe, and this is an extremely undesirable situation in a case such as ours where we want to make a statement about our particular physical system, reconstructed as a set-theoretical structure. This structure is a system of finitely many sets, so why must we go into the depths of the whole set universe in order to make a statement on such a tiny fragment of it? There is one rather effective method to mitigate the situation: We simply restrict the bound variables to appropriate scale sets over the base sets of our structure. This is indeed done in many cases of species of structures well known from mathematics. In the axioms defining a group, for instance, quantifications are restricted even to the base set of the group. But there are exceptions. Part of the usual definition of a free boolean algebra B with a set G of generators reads: Given *any* boolean algebra B' and *any* mapping h from G into B, h can be extended to become a homomorphism from B into B'. Now in this case the axiom can indeed be reduced to one where quantifications are restricted in the manner indicated. We can equivalently require that G be independent in the sense that all finite conjunctions of its elements or their complements be different from zero. However, the possibility of such restrictions certainly is not the general situation, and it seems a non-trivial problem to obtain criteria for it.

There would be no need to exert oneself in this respect if set-theoretical reconstructions of physical theories that take account of the restrictions in question were straightforward. The second introductory example I gave is indeed an easily solvable case in point. The electromagnetic potentials and their value space are submitted to existential quantification, and this can be eliminated by giving an equivalent formulation in terms of the field strengths alone. The situation is markedly different with quantum mechanics where in the usual formulation a very strong mathematical battery takes position.

It is well known that the mathematics of quantum mechanics is the mathematics of Hilbert space, but that this identification cannot be extended to the physical part of the theory. One reason for this are the complex numbers. But there is another and independent reason. It seems that no two different but proportional Hilbert vectors represent different physical states. *No* Hilbert vector, therefore, is a physical element of quantum mechanics, and Hilbert space is not even a mixed set. It is purely mathematical and with it (and the complex numbers) all its typified sets. Moreover, strictly speaking this is true for every derived structure, such as, for instance, the C^*-algebra of bounded operators or the orthocomplemented lattice of closed subspaces. In case you ever wondered what certain linear mappings or subspaces of Hilbert space have to do with observables or properties of physical systems, you were entirely justified in doing so. The best we can do in *some* cases of derived structures, like the ones mentioned, is to look at them as mathematical descriptions of certain physical structures in the sense of being *isomorphic* to the latter. Certain clearly physical structures made up of observables, states and an expectation function could be assumed to be isomorphic to certain structures derived from an infinite dimensional Hilbert space. It is in the spirit of such considerations that the last axiom in Mackey's axiomatization of quantum mechanics reads:

The partially ordered set of all questions in quantum mechanics is isomorphic to the partially ordered set of all closed subspaces of a separable, infinite dimensional Hilbert space [8].

Mackey hastens to add that this axiom is entirely *ad hoc*. Why does he do this?

The general situation before us is characterized by an axiom proper of the form

$$\bigvee Xs \in Ma. \quad \Sigma(X;s) \wedge Y \simeq P(X;s) \wedge t \simeq q(X;s). \qquad (7a)$$

Here the variables X and s are restricted to mathematical sets, Σ is a species of strutures, P and q are appropriate terms and "$\simeq$' means "isomorphic". Above all, $\langle Y;t \rangle$ is a structure about which $(7a)$ is a statement. The peculiarity of this statement calls for two comments. First, at face value the existential

---

[8] Mackey (1963), p. 71

quantification is reminiscent of cases of the first kind where coordinate systems were introduced in order to get the physical sets represented by mathematical ones. But coordinate systems can be typified and thus be made part of the structures being investigated. Although the resulting species of structures are mathematically infected by this method they will no longer suffer from quantifications going beyond the structures investigated. By contrast, in quantum mechanics no such solution seems possible. Nevertheless the problem has been attacked by other methods since 1936 when Birkhoff and von Neumann obtained a relevant result for the finite-dimensional case. Further investigations have been made by the Jauch school and recently most successfully by Günther Ludwig.[9]

Unrestricted existential quantification in $(7a)$ recedes into the background in certain other cases while the danger of mathematical overdetermination is still imminent. If in our electrodynamic example the standard representation of tensor bundles is used then the existential quantification would concern only typified sets, in which case the isomorphisms in $(7a)$ may even be replaced by equality. $(7a)$ then reduces to

$$\bigvee s \in Ma. \quad \Sigma(Y;s) \wedge t = q(V;s). \tag{7b}$$

Another example of this kind is euclidean geometry if axiomatized in terms of a distance function d. This could be criticized by pointing out the arbitrariness of fixing a unit of length. The elimination of d in favour of congruence and betweenness would be a case in point that moreover eliminates the real number set implied by d. It is true that this is another solvable case and that truly mysterious mathematical overdetermination would occur only in an unsolvable case where the strange manner in which (7) makes a statement about $\langle Y;t \rangle$ could *not* be replaced by ordinary statements composed according to the standards of (higher order) predicate logic. If there were such cases it would be very difficult to prove that there were. If, on the other hand, all physically relevant cases (7) were solvable, the solutions, although of principal interest, may turn out to be too complicated to be used in practice. The situation would still be half-mysterious in that only a mathematically roundabout route would make things acceptable to our intellectual capacity.

---

[9] To a large extent the material is included in Hooker (1975), (1979). See also Varadarajan (1968), Ch. VII; Ludwig's result is definitively presented in Ludwig (1985).

References

N. Bourbaki (1986): Elements of Mathematics: Theory of Sets. Addison Wesley: Reading, Mass.

P. W. Bridgman (1936): The Nature of Physical Theory. Princeton Univ. Press, N.J.

A. Einstein (1934, ^{2}1983): Geometrie und Erfahrung, in: Mein Weltbild, Ullstein Materialien, 119-27

H. Field (1980): Science Without Numbers. A Defence of Nominalism. Princeton Univ. Press, N.J.

C. A. Hooker (Ed.) (1975): The Logico-Algebraic Approach to Quantum Mechanics, vol.I. Reidel, Dordrecht

C. A. Hooker (Ed.) (1979): The Logico-Algebraic Approach to Quantum Mechanics, vol.II. Reidel, Dordrecht

G. Ludwig (1985): An Axiomatic Basis for Quantum Mechanics, vol.1. Springer, Berlin

G. Ludwig (^{2}1990): Die Grundstrukturen einer physikalischen Theorie. Springer, Berlin

G. W. Mackey (1963): Mathematical Foundations of Quantum Mechanics. Benjamin, New York

E. Scheibe (1986): Mathematics and Physical Axiomatization. In: Mérites et Limites des Méthodes Logiques en Philosophie. Ed. by Fondation Singer-Polignac, Librairie Philosophique J. Vrin, Paris, 251-77

E. Scheibe (1992): The Role of Mathematics in Physical Science. In: The Space of Mathematics. Philosophical, Epistemological, and Historical Explorations. Ed. by J. Echeverria et al., de Gruyter, Berlin, 141-55

P. Suppes (1960): Axiomatic Set Theory. Princeton Univ. Press, N. J.

A. Tarski (1959): What is Elementary Geometry? In: The Axiomatic Method. Ed. by L. Henkin et al. North-Holland, Amsterdam, 16-29

V. S. Varadarajan (1986): Geometry of Quantum Theory, vol. 1, Van Nostrand, Princeton, N.J.

St. Weinberg (1986): In: Mathematics: The Unifying Thread in Science. Notices of the Amer. Math. Soc. 33, 716–33

E. Wigner (1979): The Unreasonable Effectiveness of Mathematics in the Natural Sciences. In: Symmetries and Reflections, Ox Bow Press, Woodbridge, Conn., 222–37

H. Yamabe (1953): A Generalization of a Theorem of Gleason. Ann. of Math. 58, 351–65

The Mathematical Frame
of Quantum Field Theory

Klaus Fredenhagen

II. Institut für Theoretische Physik der Universität Hamburg,
Luruper Chaussee 149, D-22761 Hamburg, Germany

Since this workshop is dedicated to the role of mathematics in physics, I would like to begin with by making some remarks on this relation.

One often says that mathematics is a tool by which one is able to describe certain aspects of reality, often surprisingly well, but one should never forget that the mathematical description is not identical with reality. In principle, I agree with this statement, but in my opinion the emphasis is wrong. Actually, what we have in good theories is almost an identity between the abstract mathematical structure and the physical reality. In some sense the mathematical objects which we use are more real than the naive concept of reality which we have from our experience before we do physics. What reality is we don't know a priori, and our investigations have to correct our naive prejudices.

One could confirm this statement in several examples. I want to consider quantum field theory where a lot of advanced and abstract mathematics is successfully applied.

The aim of quantum field theory is a unifying description of particles and forces. Such a description was achieved neither in classical mechanics nor in classical electrodynamics. The back-reaction of the radiation emitted by an accelerated charged particle is a famous example for this unsatisfactory aspect of these theories. Quantum field theory can be applied to certain aspects of condensed matter physics and to elementary particle physics, and the most ambitious project is a unified description of all fundamental interactions, a project which was termed the "theory of everything". But in spite of some spectacular successes quantum field theory still is more a program for building a theory than an established theory.

There are several approaches to quantum field theory; all of them are typically very involved and mathematically sophisticated. They may be divided into the operator approach, the path integral approach and the S matrix approach. The latter had success only in special cases (integrable models in 2 dimensions). The path integral approach ignores the specific quantum aspects at the beginning and considers quantum field theory as a statistical system of classical fields where the distribution is governed by the Boltzmann factor

$\exp(iS/\hbar)$ where S is the classical action of the field theory. This approach is well suited for the discussion of semiclassical aspects (in the limit $\hbar \to \infty$ one recovers classical field theory). In its euclidean version it also can be used for a rigorous construction of some models.

The operator approach may be divided into two parts. One proceeds by postulating some algebraic relations (commutation relations or operator product expansions), constructing the associated operator algebras, and searching for suitable Hilbert space representations. The other which I want to describe in what follows starts from some general physical principles which serve to formulate a framework ("axiomatic field theory").

Despite their formal differences, all these different approaches should be considered as approaches to the same object, and a lot of logical implications are known between the different schemes.

I want to describe the so-called axiomatic approach to quantum field theory in its algebraic version. It relies on two principles, the quantum principle and the Nahwirkungsprinzip. The first one just means that one is dealing with a quantum theory; the Hilbert space formulation of quantum mechanics, however, turns out to be not well adapted to quantum field theory. The reason is that the Hilbert space formulation implicitly assumes that all self-adjoint operators in Hilbert space correspond to observables. Such an assumption is by no means required from the principles of quantum theory, and it is certainly unrealistic in a quantum field theory in infinitely extended spacetime.

Instead one considers only certain subalgebras of the algebra of all Hilbert space operators as algebras of observables. These algebras have as $C*$-algebras a simple direct characterization. They are complex algebras with an involution $A \mapsto A^*$ and a norm which satisfies $||A^*A|| = ||A||^2$. The possible outcomes of a measurement of A are the spectral values of A which can be characterized within the $C*$-algebra as those numbers λ for which $A - \lambda 1$ has an inverse. The norm of a self-adjoint operator is just the maximum modulus of spectral values.

The algebraic formulation of quantum theory offers also a solution to the measurement problem, an aspect which was first elaborated by Hepp (1972). States, as expectation functionals on the observables, are positive, normalized linear functionals on the algebra, and pure states are those which cannot be decomposed into convex combinations of different states. Now the apparent paradox of the measurement problem consists in a transition from a pure state to a mixture whose components correspond to the different possible outcomes of the measurement. In actual experiments the system under observation is not completely closed, so the time evolution is not unitary, and a paradox does not exist (see, e.g., Zeh 1993). But one may consider idealized situations where the system is closed. In such a situation one should be aware of the possibility that a sequence of pure states ω_n on a $C*$-algebra may very well converge to a mixed state ω, in the sense that the expectation values of all observables converge. One may object (see, e.g., Bell 1987) that nevertheless one will always find, for each ω_n, a suitable observable which distinguishes it from ω; mathematically, this is the fact that pointwise convergence is different from uniform convergence, and uniform convergence to a mixed state is not possible. In my opinion, this fits nicely with our experience from interference experiments in quantum theory: there are processes where, with any given finite set of observables, the interferences eventually become invisible; if one wants to observe them, one has to increase the experimental effort at every step.

We now identify a physical system with a certain C^*-algebra $\mathcal{A}$. Everywhere in physics the relation of a system to simpler subsystems is important. In a particle theory one is tempted to define subsystems by their material content. This works in certain circumstances but turns out to be unsuitable in situations where particles can be created, and it is also not practical in cases where the number of particles is very large. The basic idea underlying field theory is the Nahwirkungsprinzip according to which no actions at a distance exist. Guided by this principle one may consider subregions of spacetime as subsystems. In quantum theory one may associate with each such subregion $\mathcal{O}$ a $C*$-algebra $\mathcal{A}(\mathcal{O})$, and this subsystem is related to the full system just by an embedding of $\mathcal{A}(\mathcal{O})$ as a subalgebra of $\mathcal{A}$, such that inclusions of regions are respected.

It is the starting point of algebraic quantum field theory ("local quantum physics") to claim that the system of subalgebras $\mathcal{A}(\mathcal{O})$ fixes the theory completely, including its interpretation in terms of experiments. An additional identification of elements of $\mathcal{A}$ with concrete experiments is not necessary.

Let me briefly describe how this interpretation works in a standard case (Araki, Haag 1967; Buchholz 1987). We look at a translation -invariant system, i.e., we have an action of the translation group by automorphisms α_x of $\mathcal{A}$ such that $\alpha_x(\mathcal{A}(\mathcal{O})) = \mathcal{A}(\mathcal{O} + \S)$. We consider a state ω_0 (the vacuum) which is a ground state with respect to time translations, i.e., for all future oriented timelike unit vectors e the function $t \mapsto \omega_0(A\alpha_{te}(B))$, $A, B \in \mathcal{A}$, is a boundary value of a holomorphic functon on the upper half plane. We thus search for elements $C \in \mathcal{A}$ which may be interpreted as particle counters. C should be positive with vanishing vacuum expectation value (it should not click in the vacuum) and it should be as well localized as possible. As a consequence of the Reeh-Schlieder theorem (Reeh, Schlieder 1961), there are no positive elements with vanishing expectation values in the local algebras $\mathcal{A}(\mathcal{O})$. But we can find plenty of them which are almost local, in the sense that they can be fast approximated by local operators, i.e., for each $R > 0$ there is some $C_R \in \mathcal{A}(\mathcal{O}_R)$ such that $R^N \|C - C_R\| \to 0$ for all N, where $\mathcal{O}_R$ is the double cone with radius R centered at the origin,

$$\mathcal{O}_R = \{x, |x^0| + |\boldsymbol{x}| < R\}.$$

A coincidence experiment is described by an arrangement of counters which are spacelike separated, a notion which becomes precise at asymptotic times. We consider different future oriented timelike unit vectors $e_1, \ldots, e_n$. Then we analyze the particle content of a state ω by studying the behaviour of expectation values

$$\omega(\alpha_{te_1}(C_1)\cdots\alpha_{te_n}(C_n)) =: F(t)$$

for $t \to \pm\infty$. If for instance ω describes an outgoing particle configuration with less than n particles the expectation values tend fast to zero. If we have exactly n scalar identical particles with wave functions $\Phi_1,\ldots,\Phi_n$ we find

$$F(t) \sim t^{-3n}|\Phi_1(me_1)\cdots\Phi_n(me_n)|^2 E_1(me_1)\cdots E_n(me_n)$$

where $E_i(p) = \langle p|C_i|p\rangle$ is the sensitivity of the detector C_i for a particle with momentum p.

There are cases which are less well understood but it seems to be generally true that the theory is alredy fixed by the system of algebras $\mathcal{A}(\mathcal{O})$.

I would like to end by presenting a further example for the surprising parallelism between mathematics and physics. A basic theorem in quantum field theory is the Reeh-Schlieder theorem, which, as a consequence of locality and positivity of energy, says that the vacuum vector Ω is cyclic and separating for each local algebra $\mathcal{A}(\mathcal{O})$, i.e., $A\Omega = 0, A \in \mathcal{A}(\mathcal{O})$ implies $A = 0$, and the set of vectors $A\Omega, A \in \mathcal{A}(\mathcal{O})$ is dense in the set of vectors $A\Omega, A \in \mathcal{A}$. Now, in the theory of von Neumann algebras, there is a theorem due to Tomita and Takesaki (Takesaki 1970) that whenever a von Neumann algebra has a cyclic and separating vector Ω it also has a one-parameter group of automorphisms σ_t (the modular automorphisms) such that the state induced by Ω satisfies the KMS condition, i.e., it is an equilibrium state with a certain temperature if σ_t is interpreted as the time evolution.

In statistical mechanics this theorem has an obvious physical interpretation, and actually the historical evolution of the mathematical theory was strongly influenced by the simultaneous discussion of the KMS condition in physics (Haag et al. 1967). But for quantum field theory a physical interpretation of the theorem seemed not to be possible. But then Bisognano and Wichmann (1975) determined the modular automorphisms for the case of the wedge-like region $W = \{x, |x^0| < x^1\}$ and showed that the modular automorphisms coincide with Lorentz boosts in the 1-direction. At the same time, Hawking (1975) had found his result on the temperature radiation of black holes, and Unruh (1976) observed that a uniformly accelerated observer in Minkowski space sees the vacuum as an equilibrium state with a certain temperature. In view of the above result of Bisognano and Wichmann this result is just what one expects, since for a uniformly accelerated observer the proper time evolution coincides with the Lorentz boosts.

References

R. Haag: Local Quantum Physics. Berlin: Springer 1992

K. Hepp: Quantum theory of measurements and macroscopic observables. Helv. Phys. Acta **45**, 237 (1972)

H. D. Zeh: There are no quantum jumps, nor are there particles! Phys. Lett. **A 172**, 189 (1993)

J. S. Bell: Collected papers in quantum mechanics. Speakable and unspeakable in quantum mechanics. Cambridge Univ. Press 1987

H. Araki, R. Haag: Collision cross sections in terms of local observables. Comm. Math. Phys. **4**, 77 (1967)

D. Buchholz: On particles, infraparticles and the problem of asymptotic completeness. Proc. IAMP Conf. Marseille. Singapore: World Scientific 1987

H. Reeh, S. Schlieder: Bemerkungen zur Unitäräquivalenz von lorentzinvarianten Feldern. Nuovo Cim. X **22**, 1051 (1961)

M Takesaki: Tomita's theory of modular Hilbert algebras and its application. Lecture Notes in Mathematics Vol. 128. Berlin: Springer 1970

R. Haag, N. M. Hugenholtz, M. Winnink: On the equilibrium states in quantum statistical mechanics. Comm. Math. Phys. **5**, 215 (1967)

J.J. Bisognano, E.H. Wichmann: On the duality condition for a hermitean scalar field, J. Math. Phys. **16**, 985 (1975)

S. W. Hawking: Particle creation by black holes. Comm. Math. Phys. **43**, 199 (1975)

W. G. Unruh: Notes on black hole evaporation. Phus. Rev. **D14**, 870 (1976)

The Role of Mathematics in Contemporary Theoretical Physics

Gernot Münster

Institut für Theoretische Physik I Westfälische Wilhelms-Universität
Wilhelm-Klemm-Str. 9, D-48149 Münster, Germany

1. Introduction

One of the subjects being discussed at this workshop is *the role of mathematics in theoretical physics*. As a physicist I have an idea of what theoretical physics is. Looking at it from my own field of research, I may be allowed in the present context to restrict myself to that part of theoretical physics which aims at a theoretical description of the basic elements of the physical world: elementary particles and their interactions.

The physicist's view of mathematics, on the other hand, is always very subjective and biased. His own experience provides him only with a small window to contemporary mathematics. Therefore, when speaking about the role of mathematics in physics, I will certainly raise more questions than give answers. On the other hand, we are invited to have an open discussion, which is not limited to safe statements. So I feel encouraged to talk also about things which are outside my competence, although it may be dangerous. Concerning mathematics in general I have to rely, apart from my own impressions, on a random selection of authors who have written about developments, trends and concepts in past and present mathematics.

2. Mathematics and Physics

Let me start the discussion with the characteristic distinction between mathematics and physics. Physics is a science which aims at an understanding of the regularities in the observed world. Concerning mathematics, on the other hand, I may cite from Immanuel Kant, *Kritik der reinen Vernunft*, II. Teil: "Die philosophische Erkenntnis ist die Vernunfterkenntnis aus Begriffen, die mathematische aus der Konstruktion der Begriffe." Mathematics does not deal with phenomena and objects from the physical world, but with intellectual constructs and their relations. It is therefore counted as "Geisteswissenschaft" instead of "Naturwissenschaft". Its objects are creations of the human mind.

However, we know that mathematics is the language of physics, as has been expressed by Galileo, Kepler, and many others later. This leads us directly to the central topic of this workshop. Mathematical concepts apply to physics. How is this possible? And how did it come about? A look into history may be appropriate here.

3. The Relation Between Mathematics and Physics in History

Here is one of the places where I walk on thin ice. Being neither mathematician nor historian I have to rely on a small selection of texts on the history of mathematics (Bell 1945, Smith 1958, Meschkowski 1978, Bochner1966). The view presented will be oversimplified, but it is meant to reflect general trends and features, even if it does not apply to every single case.

The beginning of modern mathematics is often located in the 16th/17th century. If we look into the hall of fame of modern mathematics we find among others the names of Newton, Leibniz, Euler, Bernoulli, d'Alembert, Lagrange, Legendre, Fermat, Laplace, Pascal, Hamilton, and Gauss, who are equally well known as great physicists. This coincidence relates to the origin of many branches of modern mathematics. History reveals that many objects and concepts of modern mathematics have their origin in physics. Although counterexamples may certainly be found, this statement is meant to be characteristic for the rise of modern mathematics. In order to clarify it I shall give some examples.

i) Whereas Leibniz was more motivated by geometrical problems, Newton came to invent calculus from the desire to have an adequate mathematical instrument for the description of the motion of physical bodies. Under the hands of Euler, Lagrange, Hamilton, and others mechanics developed into analytical mechanics. Analytical mechanics in turn was the origin of the calculus of variations, founded by Lagrange.

ii) The Hamilton-Jacobi formulation of mechanics was one of the main origins of the theory of partial differential equations. Considerations about partial differential equations led Sophus Lie to the concept of continuous groups.

iii) Modern differential geometry has different origins. One of them is related to Gauss' duty to perform geodesic measurements in the city-state of Hannover, and to his studies of planetary motion. In this context he developed the theory of curves and surfaces of the 17th and 18th century into classical differential geometry. On the other hand there is the foundation of tensor calculus by Ricci and Levi-Civita, which was partly motivated by the use of generalized coordinates in analytical mechanics. Elasticity theory also contributed here.

iv) The investigation of the nature of orbits of classical dynamical systems by Poincaré gave rise to the foundation of modern topology by him.

v) In the area of functional analysis we may also observe various physical roots. The theory of differential equations, which was advanced by Euler and others, was motivated to a large extent by physical problems. A particular differential equation, the heat equation, led to the theory of Fourier series, which in turn is one of the starting points of Hilbert space theory. Another root of Hilbert spaces and Banach spaces is the theory of integral equations (Fredholm, Hilbert, Schmidt), which emerged from physical and technical problems.

In the early phase of modern mathematics, to which the examples refer, there was a close relation between mathematics and physics, as the above mentioned personal union of great mathematicians and physicists shows. Mathematics developed in direct contact with theoretical physics of that time.

This started to change in the following historical phase. The characteristic element of the subsequent development of mathematics is *abstraction*. With the beginning of the 19th century mathematics became more and more independent of mechanics and then physics. Mathematics began to undertake abstractions as a major element of its development (Bochner 1966). This change is not meant to have occurred to mathematics in total, but to different branches at different times according to the state of their historical development. Analytic geometry, via linear algebra, evolved into spectral theory; the theory of algebraic equations gave rise to abstract group theory and modern general algebra; algebraic geometry, topological K-theory, commutative algebra, and other abstract fields of mathematics arose, to give a few keywords only.

This process of abstraction is a natural and essential element of the way of mathematics. Yuri Manin writes (Manin 1982): "... mathematics associates to some important physical abstractions (models) its own mental constructions, which are far removed from the direct impressions of experience and physical experiment." The role of abstraction became particularly evident in our century. According to Dieudonné (Dieudonné 1978) the main concern of mathematics since 1920 has been the study of structures rather than objects. This shows up in the importance of concepts like invariants and categories. The invention of structures became a main part of mathematical activities. Many examples supporting this view can be found in Dieudonné's article (Dieudonné 1978).

The tendency towards abstraction in mathematics of course had its effect on the relation between mathematics and physics. Whereas Manin summarizes it by the statement "in this century mathematics and physics have both engaged in introspection and internal development", Freeman Dyson does not hesitate to speak of the divorce of mathematics and physics at the turn of the 19th to the 20th century (Dyson 1972).

What is the further evolution of the relation between these two sciences? A new convergence could be expected, if Courant's (Courant 1964) characterization of the general development of a mathematical discipline is correct:

"Generally speaking, such a development will start from the 'concrete' ground, then discard ballast by abstraction and rise to the lofty layers of thin air where navigation and observation are easy; after this flight comes the crucial test of landing and reaching specific goals in the newly surveyed low plains of individual 'reality'."

Before I turn to this question directly, I would like to make a digression to Wigner's thesis, which is related to it.

4. The Unreasonable Effectiveness of Mathematics in the Natural Sciences

This is the title of a much cited article by Eugene Wigner (Wigner 1960). In it Wigner points out the physicist's observation that mathematical concepts turn up in entirely unexpected connections and then often permit an unexpectedly close and accurate description of phenomena. Here he refers to advanced mathematical concepts, which are abstract and are not directly suggested by the actual world. Wigner considers them not to be simple, but to be made by the mathematicians in order to allow "brilliant manipulations". What he calls the unreasonable effectiveness of such concepts means that they apply very efficiently in physics, for example.

From the various examples given in the article (Wigner 1960) I would like to mention only the case of quantum mechanics. The theory of Hilbert space and linear operators is an abstraction, which has its roots in places which have nothing to do with quantum phenomena. A priori there is no reason to expect that interference patterns produced on a screen by streams of matter have anything to do with vectors in a Hilbert space. Nevertheless it turned out that quantum phenomena are most adequately described in terms of the mathematics of Hilbert spaces. Moreover this mathematical structure did not remain restricted to the physics of atomic spectra, where it was introduced, but turned out to be of such general validity that it governs the physics of molecules, nuclei, solid bodies, stars, etc..

The last mentioned aspect is another related point made by Wigner: the simplicity of the basic laws goes along with a large range of applicability. In the same spirit Hertz wrote about Maxwell's equations (cited in Dyson 1964): "One cannot escape the feeling that these mathematical formulae have an independent existence and an intelligence of their own, that they are wiser than we are, wiser even than their discoverers, that we get more out of them than was originally put into them", and Dyson writes (Dyson 1964): "As often happens in physics, a theory that had been based on some general mathematical arguments combined with few experimental facts turned out to predict innumerable further experimental results with unfailing and uncanny accuracy."

According to Wigner the enormous usefulness of mathematics is unexpected inasmuch as there is no rational explanation for it. From an epistemological point of view the quest for a rational explanation is of central importance. One possibility is suggested by history as discussed above: the mathematical concepts under consideration are just made so as to fit physics. Ernst Mach has introduced the term "Denkökonomie" to describe this relation.[1] This means that theories in the natural sciences are made so as to yield a highly economical description of phenomena. In some cases this may be true. But there are arguments against such an explanation: 1) in the examples discussed by Wigner and Dyson the mathematics was prefabricated for different purposes, 2) an economy of description of present knowledge cannot be expected to yield such a big success for future predictions, 3) the really surprising fact is the existence of such highly economical descriptions. Newton's law of gravitation for example was derived from rather crude observations, but nowadays we know that it holds extremely accurately and universally.

Another attempt to explain the effectiveness of mathematics, expressed by Niels Bohr, points out that mathematics, as a science of structures, provides us in principle with all possible structures. Therefore it may not be surprising that the structures adequate for physics are present in mathematics. This argument does, however, not appear convincing to me. Mathematics provides us potentially with all possible structures, but it does not so actually. The structures studied by mathematicians are certainly only a tiny fraction of all possible ones, but the ones most useful to physicists are often just among those already existing, according to Wigner.

This leads me to a point where I would like to depart a little bit from Wigner's arguments. Wigner considers the abstract mathematical structures which later turn out to be useful in physics as to be not simple but sophisticated constructs, which are chosen by the mathematicians because they allow them to exercise nice mental acrobatics. But on the other hand, considering the process of abstraction and the emergence of new concepts in mathematics, it appears that these new concepts are often directly and naturally suggested by the existing formalism. The existing formalism often intrinsically points out the ways to surmount it. This can be seen clearly in the cases of extensions of a number system. The ideal numbers, for example, were introduced by Kummer in order to uphold the unique decomposition into primes in cyclotomic number fields. Abstraction then led to the concept of ideals, introduced by Kronecker, which is a central element in abstract algebra.

In this sense, extensions of the mathematical formalism are in many cases natural and far from arbitrary.

---

[1] Mach (1897), p.5,6

5. The Recent Situation

Wigner's thesis concerning the relation between mathematics and physics is also predictive. If we subscribe to it we should expect to find more examples of mathematical concepts with unexpected applications in physics since the 1960s. In this section I would like to take a look at the recent history of the two fields in view of their relation. There are two aspects which appear to be important for the recent developments: unification, on the one hand, and the renewed relation between mathematics and physics on the other hand.

Unification. In physics one observes an increasing tendency towards unification concerning both its subject and its methods. The trend to unification in the description of physical phenomena is well known. Quantum field theory established itself as the general universal framework for the description of matter and forces. Moreover, the known particles and interactions apart from gravity can be fitted into a unified theory, the so-called Standard Model, which contains quantum chromodynamics and the Glashow-Weinberg-Salam theory. The unifying principle underlying the interactions, including gravity, is the principle of local gauge invariance. Using this principle, attempts towards even more unified theories, such as Grand Unified Theories and supersymmetric theories, are being made.

Concerning the methods of theoretical physics a certain amount of unification can also be perceived. In particular, statistical physics and elementary particle physics have experienced a big methodological overlap. I shall only mention the keywords functional integrals, renormalization group, and scaling in this connection.

In mathematics Manin (Manin 1982) also sees unifying tendencies. The theory of operator algebras, group theory, differential geometry, topology, and algebraic geometry are some of the areas whose interrelation and mutual influence seem to grow and lead to more unifications.

On the other hand, a renewed interest has also appeared in specific concrete objects. The study of imaginary quadratic number fields, L-series, and transcendental numbers are just a few examples.

The driving forces for unification are of course different in mathematics and in physics. In mathematics it is the desire to find unifying concepts and structures, whereas in physics one reaches for a unified description of fundamental phenomena in terms of theoretical models.

Relation between mathematics and physics. As has been mentioned above, theories with a local gauge invariance have played a prominent role in theoretical physics. Looking into the literature of the past twenty years one finds that there are various mathematical fields and concepts relevant for the treatment of the physics of gauge fields. Differential geometry and fibre bundles turned out to represent the proper mathematical framework for gauge field theories; topology and characteristic classes played an important role in connection with instantons, and the physics of anomalies found its adequate description in terms of cohomology theory, index theorems and related items, to mention some examples.

This is clearly another instance of Wigner's observation. Differential geometry, in particular principal fibre bundles, Cartan-Ehresmann connections, etc., were developed on the basis of mathematical reasoning which had nothing to do with non-Abelian gauge theories, which were only introduced into theoretical physics in the late 1950s. Nevertheless they appear to yield the mathematical structures which fit perfectly well the needs of gauge theories.

Conversely, there is an interesting feedback from physics to mathematics in recent history. Considerations about instantons, which are solutions of the gauge field equations, have led to astonishing new results about the geometry of four-dimensional manifolds. Conformally invariant quantum field theories and field theories with Chern-Simons action have brought out new fruitful approaches to knot theory and to the topology of three-dimensional manifolds. Even more fascinating in this respect is the case of string theory, which employs topics as different as Riemann surfaces, algebraic geometry, lattices, infinite-dimensional Lie algebras, modular forms, number theory and others, and has revealed new connections between these fields, one example of which is known among mathematicians as "monstrous moonshine".

In accordance with the view presented above, Manin (Manin 1982) finds that the present relation between mathematics and physics is characterized by an increasing overlap between them, and he observes a growing willingness to learn from each other and to transfer tools and techniques. With respect to Wigner's thesis I think recent developments are in support of it in even more unexpected circumstances.

In closing I would like to cite Courant again (Courant 1964): "That mathematics, an emanation of the human mind, should prove so effective for the description and understanding of the physical world is a challenging fact that has rightly attracted the concern of philosophers."

References

Bell, E.T. (1945): The development of mathematics, McGraw-Hill.

Bochner, S. (1966): The role of mathematics in the rise of science, Princeton University Press.

Courant, R. (1964): Mathematics in the Modern World, Scientific American, September 1964, p. 41.

Dieudonné, J. (1978): Present Trends in Pure Mathematics, Advances in Mathematics **27** 235.

Dyson, F.J. (1964): Mathematics in the Physical Sciences, Scientific American, September 1964, p. 129.

Dyson, F.J. (1972): Missed Opportunities, Bull. Am. Math. Soc. **78** 635.

Mach, E. (1897): Die Mechanik in ihrer Entwickelung historisch-kritisch dargestellt, F.A. Brockhaus, Leipzig.

Manin, Yu.I. (1982): Mathematics and Physics, Birkhäuser. Boston.

Meschkowski, H. (1978): Problemgeschichte der neueren Mathematik, B.I. Mannheim.

Smith, D.E. (1958): History of mathematics, Vol. I,II, Dover Publications.

Wigner, E.P. (1960): The unreasonable effectiveness of mathematics in the natural sciences, Commun. Pure and Appl. Math. **13** 1.

A Most General Principle of Invariance

Erhard Scheibe

Am Moorbirkenkamp 2a, D-22391 Hamburg, Germany

I

The subject of invariance or symmetry that I am going to talk about is an interesting subject for various reasons. To the philosopher of science it is particularly interesting because here he finds the physicist saying things that he usually does not say: he finds him making certain metastatements of a nonempirical status. Put a physicist before the work of Carnap and he will shrug his shoulders. He is more than happy, however, to state that the laws of electrodynamics are invariant under Lorentz transformations or to postulate the relativistic invariance of any future law. But in doing this the physicist does exactly the kind of thing that Carnap wanted a philosopher of science to do: he states or postulates something about a physical law. Moreover, his claims might even be of a purely syntactical nature: frequently one hears it said or reads it in print that it is the form of a law that is invariant. It seems, therefore, that invariance, if anything, is a subject of common interest to the philosopher and physicist, nicely suited to be dealt with in a meeting like this.

The metatheoretical status of an invariance statement is part of what might be called 'Wigner's hierarchy'.[1] By 'the hierarchy of our knowledge', as he himself calls it, Wigner means "the progression from events to laws of nature, and from laws of nature to symmetry or invariance principles". Speaking of a progression Wigner obviously has in mind that just as by physical laws we make statements about events so by invariance principles we make statements about physical laws, thus progressing to a higher abstraction. Wigner[2] even sees "a great similarity between the relation of the laws of nature to the events on one hand, and the relation of symmetry principles to the laws of nature on the other". The similarity is that just as "if we had a complete knowledge of all events in the world ... there would be no use for the laws of physics" so similarly "if we knew all the laws of nature ... the invariance properties

---

[1] Wigner (1979), 30 f.

[2] Loc. cit., 16 f.

of these laws would not furnish us new information." This is but another way of saying that invariance statements are analytical, and again it seems a curious observation to see physicists being fond of making such statements. In a recent textbook on particle physics we read[3]: "It is no exaggeration to say that symmetries are the most fundamental explanation for the way things behave (the laws of physics)." In view of such a statement one is inclined to ask whether we are right after all in teaching students that analytical statements don't tell us anything. Moreover, it seems that I need not beg your pardon for drawing your attention to a principle of invariance that on account of its extreme generality has little chance of having any physical meaning at all.

Still by way of introduction let me illustrate the typical situation in which a physicist makes an invariance statement by the following geometrical example. Suppose our theory is about figures in Euclidean space, and the basic law of the theory simply states that a figure is a sphere. Then a valid invariance statement is that this law is invariant under Euclidean transformations: the property of any figure of being a sphere obviously is preserved under all Euclidean transformations. In the numerous books on symmetry in physics that we owe to present fashion[4] the authors usually start with the even more concrete and pictorial invariance of a singular figure, e.g., a sphere as being invariant under all rotations around its centre. However, I may be allowed to start with the more abstract situation in which the question is about the invariance of a certain property of a figure and not the figure itself.

Even in such a case, which is nearer to that of a physical law in the proper sense, the invariance statement is fairly specific. Our statement that the property of being a sphere is invariant under Euclidean transformations has the overtone that this property is *not* invariant under all differentiable transformations of space and not even under all affine transformations. Thus it seems that to a given invariance statement statements of non-invariance with respect to the same law are readily at hand, pointing out the specificity of the former. However, the statement that the property of being a sphere is not invariant under all affine transformations is a rather peculiar statement. It is true only under the tacit assumption that our statements about figures before and after the transformation refer to the *same* Euclidean metric. The same observation holds, of course, for the positive statement of Euclidean invariance. However, as soon as we admit the very possibility of submitting also our metric to the transformations in question there is a significant difference between the two cases: whereas the Euclidean metric is preserved under Euclidean transformations it is *not* preserved under all affine transformations. Therefore, in the second case two possibilities open up: either we stick to the original metric or we submit the metric to the very same transformations by which the sphere is transformed. The point now is that, whereas in the former case we get the statement of non-invariance as mentioned, in the latter case *we regain invari-*

---

[3] Dodd (1984), 41.
[4] See, for instance, Genz und Decker (1991); Mayer-Kuckuk (1989).

ance: the affinely distorted sphere becomes again a sphere with respect to the affinely distorted metric.

We thus see that the apparent specificity of our original invariance statement is brought about by the concurrence of two factors: 1) a more general and, as we shall see in a minute, an even excessively general invariance – in our case the invariance of the axioms, fixing the idea of a Euclidean space with a sphere distinguished in it, with respect to a very general class of transformations, and 2) a special case of this invariance characterized by a restriction to those transformations leaving invariant a given common fragment of the physical entities considered by the theory – in our example the *one* Euclidean space of Newtonian physics. According to this two-fold explanation the theory of invariance to be developed in the following views a physical theory as being a pair consisting of *axioms* as usual and a *frame* in which the axioms are interpreted. In the axioms we talk about a physical system, and the frame is some fundamental part of the system ranking on the level of the theory – like Euclidean space in our theory of spheres. More specifically, we may think of the axioms as being extensions of set theory talking about structures describing physical systems, the frame being itself a particular structure. Then the transformations with respect to which we look for invariances may be arbitrary isomorphisms of structures, and the entities, being invariant, may be either statements about structures or structures themselves. Under these assumptions the most general principle of invariance mentioned in the title of the paper and already alluded to in the given illustration says that the axioms of a physical theory are invariant under arbitrary isomorphisms. However, since it is only the frame of a theory that gives it a physical meaning, only those isomorphisms leaving the frame invariant are of physical interest.

Thus the point of the following analysis is that invariance has two aspects: an unconditional and a conditional one. Under the unconditional aspect an invariance statement is viewed as part of the extremely general invariance principle that, put in terms of mathematical logic, any structure isomorphic to a model of a theory is itself a model of that theory. In other words; you may say about a structure almost anything, what you say, if true, is also true of any structure isomorphic to the first one, and if false then false. *But* – and here comes the other aspect – invariance statements usually are conditioned by referring only to those isomorphisms that leave invariant some fundamental structure needed for the interpretation of the theory. It is this condition that makes the invariance statement a more or less specific one, leading to names like Galilean invariance, Lorentz invariance, etc. The reason for this specificity is 1) the specificity of the distinguished structure – the frame as I called it – and 2) the fact that, whereas any isomorphism leaves invariant almost any *statement* about structures, only very few isomorphisms leave a given *structure* invariant. This double-aspect theory of invariance seems to hold in physics without exception as far as it goes, and it certainly covers the field to a considerable extent. However, not all transformations with respect to which interesting invariance statements have been made in physics are isomorphisms of the physical systems treated by a theory. Therefore, in due course a generalization of our approach has to be indicated.

II

After this introductory overview I now come to some details. In the second part of the paper they are details of the general theory of invariance to be proposed. In the first place a word on the concept of physical theory is in order. According to the double-aspect of unconditional and conditional invariance we have to fix our eyes on two parts of a theory: its formal axioms and its meaning-generative frame. For it it the axioms that show the unconditional general invariance, and it is the frame whose own invariance leads to a conditional invariance of the physical laws (as part of the axioms).

Let us first address ourselves to *unconditional invariance* and therewith to the *axioms* of a theory. Then two things are important for us: 1) that the axioms say something about some physical system, e.g., a field or a system of particles, and 2) that we may view the physical system as a structure in the sense of modern mathematics and mathematical logic. What we wish to say about a physical system we can then formulate as a statement about a structure[5] using some codification of set theory as our language and logic (in the formal sense). The axiom of our physical theory thus assumes the form

$$\Sigma(X;s) \equiv s \in \sigma(X) \wedge \alpha(X;s). \tag{1}$$

Here X and s each stand for a finite system of sets, such that $\langle X;s \rangle$ makes up a set-theoretical structure. This is expressed by the first member of the right-hand side in (1) saying that the sets s are elements of scale sets over the sets X, i.e., of sets generated from the X by successive formation of power sets and cartesian products. The second member α in (1) is the axiom proper, and thus (1) symbolizes a species of structures for each σ and α. Simple examples, e.g., groups and topological spaces, are well known from mathematics. For us, however, the point is that by their appropriate combination species of structures can be formed that may be used directly as the axioms of a physical theory[6].

Now in the abstract treatment of structural mathematics as we find it, for instance, in Bourbaki's encyclopedia a well determined property of the axiom proper α in (1) is presupposed or even explicitly required. And this property is an invariance property – *canonical invariance* as I will call it[7]. It is an invariance of α under arbitrary isomorphisms of the structure about which α is

---

[5] For details see Bourbaki (1968), Ch. IV.

[6] The first to have applied species of structures to physics in a systematic fashion is Ludwig in (1978, [2]1990).

[7] In Bourbaki, loc. cit., Ch. IV, § 1 the term for 'canonical invariance' is transportability.

a statement. Let me first briefly recapitulate what an isomorphism is in this context. If we consider any bijections of the principal base sets X onto sets X' then bijections of every scale set over the X onto the corresponding scale set *of the same type* σ over the X' are canonically induced. In particular, any structure $\langle X; s \rangle$ is mapped onto an *isomorphic* structure $\langle X'; s' \rangle$. Without giving the formal definitions (which are straightforward) I would like to emphasize that these canonical extensions or representations of originally given bijections are in no way dependent on the species to which a structure belongs. Such a dependence might be suggested by terms like homeomorphism, diffeomorphism, and the like as being isomorphisms of topological spaces, manifolds, etc., respectively. However, given, say, a group we do not have to consider the group axioms in order to construct any isomorphism of the group onto another structure. The real phenomenon to be observed is that, although the isomorphism can be chosen completely independently of the group axioms, the structure to which it leads *is again a group*.

This is is not quite the most general situation, though. According to its construction the typification indeed *is* invariant, i.e., we always have

$$s \in \sigma(X) \leftrightarrow s' \in \sigma(X'), \tag{2a}$$

where the scale term σ is the *same* on both sides of this equivalence. We cannot expect the same to hold for the axiom proper without exception. However, it is remarkable that the *requirement* that also

$$\alpha(X; s) \leftrightarrow \alpha(X'; s'), \tag{2b}$$

with the same α on both sides holds for every isomorphism is satisfied for any given theory of physics. We shall see in a minute that this invariance in all its generality may not enjoy life to the full for *interpretative* reasons. Interpretations reduce symmetries. But this does not alter the fact that the physical axioms *are* canonically invariant as regards their *form*[8].

Can we understand why they are? I have no satisfactory answer. Canonical invariance somehow expresses that the axioms don't tell us anything about the nature of the elements of the principal base sets – that these can be chosen or 'interpreted' quite arbitrarily. We cannot, for instance, require that two of the principal sets have a non-empty intersection or that one of them is a particular set without violating our principle (2b). The first case could be a relative and the second an absolute determination of the elements in question. By contrast, for the typified sets s it is uniquely determined what their elements are, once the principal base sets X are fixed. Perhaps one could say that canonical invariance is a very weak condition of *lawlikeness* of the physical axioms, and this would be desirable at any rate. All further considerations apparently must refer to concrete examples for the time being.

Let me now come to the other part of a physical theory in which the aspect of *conditional invariance* is linked to the concept of the *frame* of a

---

[8] See Scheibe (1981), 311–331.

theory. Up to this point our consideration was essentially about formulas. We now simulate their physical interpretation by a fixed frame structure that is a common fragment of those structures $\langle X; s \rangle$ that our theory axiom (1) deals with. The frame of a theory thus becomes a common part also of the physical system to which the theory refers indeterminately. It gives a theory meaning without fixing a reference, to use the Fregean terms. The separate development of theoretical physics beside experimental physics is in need of such a distinction anyway. The talk of space and time, of particular quantities like momentum and energy, temperature and entropy, electric and magnetic fields, etc., has meaning already within theoretical physics *without* any real physical system being fixed thereby.

We can take account of the frame structure by means of a decomposition

$$\Sigma_0(X_0; s_0) \wedge s \in \sigma(X_0) \wedge \alpha(X_0; s_0, s) \tag{3}$$

of Σ in (1) where $\langle X_0; s_0 \rangle$ is the frame. Side by side with the axioms the frame is part of the theory such that its change would change the theory, too. By contrast, the set s is still variable within the theory and indicates the many physical systems the possibility of which is stated in (3). Paradigm cases of frame structures in Newtonian physics are, of course, space and time. They are so because of their unique existence. It is hard to see how this can be taken into consideration other than by making these very structures themselves part of the theory. For even if we are in the possession of a categorical theory of, for instance, Euclidean space, our subject is fixed *only* up to an isomorphism. And we *cannot* by any means of ordinary axiomatics reduce this multiplicity. We cannot do so precisely because of canonical invariance.

Given a frame structure as part of a physical theory it is obvious that the invariance situation changes from an unconditional to a conditional one. Formerly all isomorphisms were to be admitted. Now those isomorphisms leaving invariant the frame structure are distinguished, and whatever *their* meaning may be it is hard to see what meaning could be attached to isomorphisms *not* leaving invariant the frame. In this way we obtain the invariance

$$\alpha(X_0; s_0, s) \leftrightarrow \alpha(X_0; s_0, s') \tag{4a}$$

as a consequence of unconditional canonical invariance now under the condition that

$$X_0' = X_0, \quad s_0' = s_0. \tag{4b}$$

It is these conditional invariances, concerning only s, that we typically meet with in the textbooks as the classical invariances, like the Galilean or Lorentz invariance of a physical law. But we now see that as *invariances* they are nothing but special cases of canonical invariance *following* from our general principle together with the special condition that in one case our frame structure is Galilean spacetime, in another case Minkowski spacetime, etc. Once we have decided about the frame no further invariance postulate is needed for any given law.

III

After these general considerations it is now time to look at some examples. Although not all invariances to be found in physics are canonical, sufficiently many canonical invariances are left to make the demonstration of their extensive presence in physics somewhat laborious. The following selection is by no means representative. But it may serve to draw our attention to some issues that may easily lead to misunderstandings and later on also to some more serious questions.

Let us first look at the unitary invariance of the Schrödinger equation

$$i\dot{\psi} = H\psi \qquad (\hbar = 1). \tag{5a}$$

The term 'unitary' signalizes that we take the quantum theoretical state space S – a Hilbert space – to be our frame. Precisely if we do this we can restrict further consideration to the canonical representations of automorphisms of S, i.e., of unitary transformations. For unitary U the usual formulation of the corresponding transformations of the time development $\psi(t)$ of the states and of the Hamiltonian operator H is

$$\psi'(t) = U\psi(t), \quad H' = UHU^{-1}. \tag{5b}$$

In case you ever wondered why we take these representations to be the 'correct' ones the theory under discussion has the answer: because they are canonical. For the Hamiltonian operator the argument is that its typification in the state space S is

$$H \in \mathrm{Pow}(S^2).$$

The canonical representation of U on the scale set $\mathrm{Pow}(S^2)$ then already yields a uniquely determined image H' of H whatever H may be. If H is an operator then the transformation has the form given in (5b). Moreover, if H is linear and self-adjoint then the same follows for H'. In this way the example clearly illustrates how a canonical representation beyond its existence in general is provided with additional properties as a consequence of additional assumptions about the structures in question. The same holds for the state functions in (5b), and from both it follows as usual that (5a) is invariant.

The quantum theoretical example may evoke the question why we have chosen the state space to be the frame of the theory and consequently have distinguished the unitary group. The simplest answer is: why not? This is to say that here we have a very obvious freedom indeed. From a purely *formal* point of view we could as well have kept fixed the Hamiltonian operator or have made the metric of the state space variable. Moreover, not only is the decomposition (3) in general arbitrary. To *every* fragment $\langle X_0; s_0 \rangle$ we may assign its automorphism group and to this in turn its (relevant) canonical representation. In every such case we shall find the corresponding special canonical invariance (4). Insofar the procedure depends on nothing but a given species of structures together with a decomposition (3).

As regards the *interpretation*, however, our example allows us to observe the following: we know of numerous physically realized instances where for a given interpretation of the state space various Hamiltonian operators are applied, describing so many different interactions. But we do *not* know of anything similar for the metric of the state space vis à vis its linear structure. For a *given* interpretation the situation here is the same as we found in the case of Euclidean space. Our freedom of choice is drastically reduced for empirical reasons. Yet general quantum theory has no unique frame structure attached to it. The reason is simply that the theory *has* no unique interpretation in the sense that it could be represented by a frame. Even quantum mechanics proper, i.e., quantum theory extended by a representation of the canonical commutation relations, can be assumed to have a unique interpretation only after fixing the degree of freedom[9]. The same situation is met with in classical Hamiltonian mechanics. But one cannot blame someone for making a frame part of a physical theory by arguing that there is no unique frame to be found in these cases. For either these cases are mere formalisms or, if they are interpreted, a frame will have been attached to them.

Already in the introduction it was said that canonical invariance is not the only kind of invariance that has found the interest of physicists. Since non-canonical invariance is not the object of this paper the following example may rather be viewed as an instructive counterexample to canonical invariance than as a beginning of a generalization. The point of canonical invariance is that the solutions of a physical law are always typified sets. Therefore, transformations induced by canonical representations of transformations of those sets *by which* the former are typified certainly are among all possible transformations. But by no means they do exhaust this class. *Gauge transformations* are a case in point[10]. They concern, for instance, a quantum mechanical but relativistic particle with state function ψ moving in an electromagnetic field with potentials A where both ψ and A are defined on Minkowski space-time M. The equations of motion (that need not be given here) are then left invariant by the gauge transformations

$$\left.\begin{aligned} \psi'(x) &= e^{i\alpha(x)}\psi(x) \\ A'_\mu(x) &= A_\mu(x) - \frac{1}{e}\frac{\partial\alpha}{\partial x^\mu}, \end{aligned}\right\} \tag{6a}$$

where α is any real function on M. By contrast, the transformations of ψ and A induced by a Lorentz transformation of M would be

$$\left.\begin{aligned} \psi'(x') &= \psi(x) \\ A'_\mu(x') &= \frac{\partial x^\lambda}{\partial x'^\lambda}(x)A_\mu(x). \end{aligned}\right\} \tag{6b}$$

---

[9] This is due to the von Neumann-Stone theorem, see Emch (1984), Ch. 8.3 f

[10] See, for instance, Bethge und Schröder (1986), Ch. 5.

It is obvious how the additive group of functions α operates *directly* on the state functions while a Lorentz transformation on M has to work all the way up to the entities in whose transformations we are interested. The dichotomy between canonical and non-canonical invariance thus illustrated does not coincide with any of the usual distinctions between geometrical and non-geometrical, internal and external transformations and the like. But it is certainly more precisely definable than the latter, and it may even be more important.

IV

With my third example I come back to canonical invariance in order to discuss its relation to the well-known invariances occurring in analytical geometry. Let us take a static scalar field in Euclidean space obeying, for instance, the Laplace equation

$$\Delta\phi \equiv 0. \tag{7a}$$

In this case we are dealing with the Euclidean invariance of this equation under the transformations

$$\phi'(\boldsymbol{r}) = \phi(A^{-1}\boldsymbol{r}) \tag{7b}$$

where A is a Euclidean transformation. The statement that this case, too, is a case of canonical invariance with Euclidean space as our frame is as trivial as was the corresponding statement for the first example. However, the present case is well suited to discuss a somewhat delicate point that sometimes goes under the name of *covariance*. I take the occasion to emphasize the essential difference between canonical invariance and covariance in the sense of invariance under coordinate transformations.

The statement that the law (7a) via the representation (7b) is invariant under Euclidean transformations is ambiguous. It may mean that the transformations of $\mathbb{R}^3$ (!) leaving invariant the quadratic form

$$(x_1 - x)^2 + (y_1 - y)^2 + (z_1 - z)^2 \tag{7c}$$

also leave invariant the Laplace equation if the latter is understood to be an equation for real functions on $\mathbb{R}^3$ transforming according to (7b). But this *numerical* invariance statement, as it might be called, certainly cannot be its primary meaning as an invariance statement concerning a *physical* theory. Rather the primary meaning must be such that the numerical invariance statement is but an expression of the invariance statement proper *in a Euclidean coordinate system*.

This transpires already from the necessity to invoke an equation different from (7a) and a function different from (7c) where we express our invariance statement proper in a non-Euclidean coordinate system. This observation does not, of course, answer the question of how to give a sound formulation of the invariance statement proper in the present case. And even if an answer were produced we would still be saddled with the different question of what it means that the numerical invariance statement expresses the proper one in a Euclidean coordinate system. A precise and general clarification of this matter would be beyond the scope of this paper. The following, I hope, will clarify the situation in its essentials.

All geometries that ever have been applied in physics allow the introduction of coordinate systems. Even if a coordinate-free axiomatic is available and used, the existence of a coordinate system can be proved. The simplest procedure, however, is to introduce coordinate systems from the very beginning and to exploit fully the basic principle of analytic geometry, i.e., the principle of saying what we want to say about a physical system by using coordinate representations. In recent publications analytical geometry is dismissed, though not for the very definition of a manifold. Still, the authors try to make their presentations as coordinate-free or intrinsic as possible. Now it is true that equivalent formulations of a geometry can be given where one of them is more intrinsic than the other. At present my point only is that to the extent to which coordinate systems *are* used a new invariance phenomenon comes into play. I shall be using the term 'covariance' for this phenomenon, being aware of the fact that some authors use the term in a different sense[11]. Roughly put, covariance is invariance under coordinate transformations. If, for instance, we wish to do Euclidean geometry in an analytical fashion we start with the requirement that there be a (global) coordinate system of space in which the square of the distance is given by the quadratic form (7c). This requirement establishes the class of Euclidean coordinate systems as a distinguished class. If we now want to formulate a field law we can do so by using a representation of the field in a Euclidean coordinate system, demanding that the representation satisfies, for instance, the Laplace equation. For an *arbitrary* euclidean coordinate system this gives us a consistent condition for the field *itself* only if the equation is (numerically) invariant under all Euclidean coordinate transformations. It is this mathematical or – as I called it – numerical invariance that, in its representative role, deserves to be given a name of its own, e.g., covariance. In the first place covariance is the basis for a consistent introduction of a physical field law by a mathematical equation. So covariance is different conceptually from canonical invariance, even from its conditional version. The two are correlated, though. For the automorphism group of (7c) is simply the coordinate representation of the canonical automorphisms of the Euclidean metric in physical space.

The principal difference between canonical invariance – conditional or unconditional – and geometrical covariance becomes even more evident if we admit arbitrary differentiable coordinates in space. In this case we have to find a generally covariant formulation of the Euclidean metric and the field

---

[11] Einstein's favourite view of the matter is discussed in Scheibe (1991), 23–40.

equation. It is well known how this is done by means of the metrical coefficients g_{ik}, and it is clear how in this way on the side of coordinate representations the very large group (or pseudo-group) of differentiable transformations (in $\mathbb{R}^3$) comes into play. But all this has only to do with the coordinate representation of our theory, this theory itself still being characterized by Euclidean space and the Laplace law[12].

V

Next to the principle of canonical invariance the theory proposed in this paper rests on the notion of a frame structure as part of a physical theory. Canonical invariance is unconditional with respect to isomorphism, and its principle is *one* for the whole of physics. Different kinds of invariances do not come into play even by the fact that many *formally* inequivalent theories, i.e., theories inequivalent by their axioms, are used in physics. Apart from non-canonical invariance we meet with a multiplicity of kinds of invariance only on account of the different frame structures of the theories, i.e., their different *contents*. Consequently, we are not concerned here with a variable invariance behaviour of physical *statements* but with so many different restrictions of the one canonical invariance to the respective theory frames. This situation suggests the question to what extent the variety of theory frames can be reduced, possibly within general theory reduction, and what criteria of irreducibility, if any, are at hand. It is, of course, almost preposterous to raise such a far-reaching question at the end of this talk. Let me make some remarks in conclusion, though, that are related to the invariance business.

We have already seen the most trivial case where a frame structure is chosen to be more special than our reductive abilities would require it to be. Even after the emergence of general quantum theory the quantum mechanics of the hydrogen atom is a theory in its own right. More serious is the case where the fact that we don't reduce because we simply are at a loss to do so becomes the basis for attempts to give reasons why reduction is impossible. Euclidean space and its explanation as a pure intuition certainly are historical examples for this situation. The Euclid-Hilbert theory of space is an interesting theory on many accounts. But in matters of invariance the really important thing is the uniqueness of space. As Kant put it[13]: "We can represent ourselves only one space; and if we speak of diverse spaces, we mean thereby only parts of one and the same unique space." This, then, leads to a unique automorphism group occurring in every theory about objects in space.

The relation of Euclidean space to its theory throws light also on the next case where alternatives to a given frame are available but only as models of one

[12]For the general notion of analytic geometry see the paper mentioned in the preceding footnote.

[13]Kant, *Critique of Pure Reason*, B 39.

categorical theory. Thus, theoretically there *are* different Euclidean spaces, if only isomorphic ones. Is this situation different in principle from the variety of solutions of, say, Maxwell's equations that, of course, include an infinity of non-isomorphic ones? The answer demands a distinction. The Hilbert spaces of quantum mechanics have different empirical realisations although they are all isomorphic. However, they cannot be used as state spaces in this abstractness anyway. There is no interpreted physical theory having a state space in its frame whose elements remain entirely indeterminate. In the contrary case we would immediately ask: *what* are the states of the theory? In quantum mechanics it is only the spectral decompositions of a Hilbert space for which we can answer this question. But they are no longer isomorphic. The situation is different for Newtonian space and Minkowskian spacetime. Here the question 'what space points or what spacetime points do you mean' would leave the physicist rather speechless. The categoricity of Euclidean or Minkowskian geometry gains in importance. At present we are still prepared to say: two Euclidean spaces or two Minkowskian spacetimes, *if* they occur in a fundamental position within a frame, cannot be distinguished by physical means. We could simply not say *which* of the two spaces or spacetimes is ours. And it is for this reason that we can replace a categorical theory by one of its models. The proviso mentioned, however, is decisive. There are many isomorphic Euclidean spaces provided by the different inertial systems of Minkowskian (or Galilean) spacetime, and already Newton confronted his absolute space with variable relative and as such empirical spaces.

Difficulties come up with general relativity. The outstanding event in the transition from special to general relativity was the new contingency of the metric – a contingency reaching far beyond isomorphic models. There is no longer any question of categoricity on this level, and even topology is in the grip of this process. There is hardly anything left for a universal frame, and there is no excuse as we had it for the mechanical theories where irreducibility was not required. In general relativity no deduction is made from the fundamental position of spacetime. So I don't really know what to say in this case. One thing, however, seems to me to be no difficulty at all. There is still local categoricity on the topological level. If we take this as an occasion to choose the manifold of spacetime as the corresponding frame, our automorphism group would become the fairly large group of all diffeomorphisms of the manifold chosen. I now quote Sommerfeld who said in a popular lecture[14]: "The theory of special relativity amounts to a theory of the invariants of the Lorentz transformations ... General relativity, too, is the theory of the invariants of the natural laws. It is only that here the group of Lorentz transformations is replaced by the total group of all coordinate transformations of the four-dimensional universe." In my view this is an essentially correct description of the situation although it is given in the usual sloppy way of the physicist.

---

[14]Sommerfeld (1968), 640–643.

References

Bethge, K., Schröder, U.E. (1986): Elementarteilchen, Wissenschaftliche Buchgesellschaft: Darmstadt.
Bourbaki, N. (1968): Theory of Sets, Addison Wesley: Reading, Mass.
Dodd, J.E. (1984): The Ideas of Particle Physics, Cambridge University Press: Cambridge.
Emch, G.G. (1984): Mathematical and Conceptual Foundations of 20th-Century Physics, North-Holland: Amsterdam.
Genz, H. und Decker, R. (1991): Symmetrie und Symmetriebrechung in der Physik, Vieweg: Braunschweig.
Ludwig, G. (1978, 1990): Die Grundstrukturen einer physikalischen Theorie, Springer: Berlin.
Mayer-Kuckuk, Th. (1989): Der gebrochene Spiegel, Birkhäuser: Basel.
Scheibe, E. (1981): Invariance and Covariance, in: J. Agassi and R.S. Cohen (Eds.), Scientific Philosophy Today, Reidel: Dordrecht
Scheibe, E. (1991): Covariance and the Non-Preference of Coordinate Systems, in: G.G. Brittan (Ed.), Causality, Method, and Modality, Kluwer: Dordrecht.
Sommerfeld, A. (1968): Philosophie und Physik seit 1900, in: Gesammelte Schriften, Vol. IV, Vieweg: Braunschweig.
Wigner, E.P. (1979): Symmetries and Reflections, Ox Bow Press: Woodbridge, Conn.

Kant and the Straight Biangle[1]

Gerold Prauss

Philosophisches Seminar I, Wertmannplatz, D-79098 Freiburg,
Germany

In the framework of his critical philosophy, Kant, as is well known, felt obliged
to adopt a position with regard to the problem of the foundations of geometry.
One of the ways in which he did this was to consider whether the concept
of a straight biangle was consistent or involved a contradiction. Kant then
entangled himself in a striking difficulty, which, however, appears to have
been noted by few. To my knowledge, the only other person who has as yet
seen it is Gottfried Martin[2] and his proposed solution is also the only one that
I yet know.

This is all the more surprising, since the difficulty consists of an explicit
contradiction. For in one passage in the *Critique of Pure Reason* Kant regards
the concept of a straight biangle as free of contradiction[3] but in another as
involving a contradiction[4], and there is a further passage in which this second
opinion is supported[5]. Martin's proposed resolution is, moreover, of a kind that
Martin himself admits makes him "uncomfortable"[6]. The problem is that it is
in no way possible to regard – and this is Martin's proposal –the passages in
which Kant speaks of the contradictory nature of this concept as a remnant
of his precritical period, i.e., to include them among the texts that Kant had
drafted much earlier and only later took over into the *Critique of Pure Reason*,
doing this without noting the contradiction. All these passages undoubtedly
belong, both in themselves and in the context in which they occur, to the full
flowering of Kant's critical philosophy.

There is therefore no reason why, given the two passages that declare the
concept to be self-contradictory, one should prefer instead the single passage in
which the concept is said to be free of contradiction. Indeed, there is actually

---

[1] translated from German by Julian Barbour, Oxon, UK.

[2] *Das geradlinige Zweieck, ein offener Widerspruch in der Kritik der reinen Vernunft*, in: *Tradition und Kritik*, Festschrift für R. Zocher, W. Arnold ed., H. Zeltner, Stuttgart 1967, p. 229-235

[3] A 220 f. B 268

[4] A 291 B 348

[5] A 163 B 204

[6] see quotation above

a stronger argument for rejecting this one passage, not only on philological grounds – it is after all just one passage – but also at a basic level, because the passage ultimately betrays an uncertainty of Kant with regard to his own critical philosophy. The passage is in fact as follows: "There is thus in the concept of a figure enclosed between two straight lines no contradiction ... the impossibility resides not so much in the concept in itself as in the construction of the concept in space"[7], i.e., it resides in the fact that one cannot generate an intuition for the concept. Under such circumstances, one must ask how then this concept is to possess definiteness as meaning or sense; for in accordance with Kant's critical philosophy it can have such definiteness only together with intuition.

With everything thus speaking against this passage, Martin nevertheless adopts it, but only because he regards it as an advance. His argument is that Kant had in the meanwhile recognized that a straight biangle was admittedly impossible, but not because the concept of a "straight biangle" was self-contradictory, but because, despite being entirely consistent, it did not allow "construction", i.e., the intuition corresponding to it could not be formed[8]. This then appears as an advance because the straight biangle, being recognized as impossible only *synthetically*, is nevertheless for the first time possible *analytically*, and in this way Kant would have opened up a possibility for non-Euclidean geometry, a possibility that in the framework of his critical philosophy, however, simply had to lead to insoluble difficulties.

From the start therefore we need consider only a solution that can explain why, at the peak of his critical philosophy, Kant himself was still uncertain which choice he should make in what for him was evidently a delicate question.

As I should like to show at the end, the actual reason for this hesitation is that, to the last, Kant remained undecided about a question that touched the foundation of his entire system. In fact, there are questions here which must be addressed to the mathematicians and physicists and which are much more far reaching than has hitherto been assumed, namely, whether, from Kant's point of view, the issue of Euclidean versus non-Euclidean geometry and their relationship to each other really has been so definitively decided as has up to now been assumed. The point at issue touches the relationship between concept and intuition, doing so, moreover, in the truly fundamental case of "straight" or "straight line". For the question of whether the concept of a "straight biangle" is consistent even though the concept cannot be "constructed", namely, cannot be associated with intuition in the form of space, clearly depends on what "straight line" is to mean. Thus, Kant's difficulty in this question is also related to the difficulty of defining the concept of a "straight line", which already since Euclid and up to the present day remains unsolved. Is this problem really insoluble, i.e., only soluble through an axiomatic system as adopted by

---

[7] A 220 B 268

[8] cf. quotation above, p. 234 f.

Hilbert, in other words, only through an implicit if not an explicit definition of a straight line?

In view of the obscurity of its definition already in Euclid[9], geometers withdrew more and more, at the latest from the time of Archimedes[10], to the position that a straight line was to be regarded only as the shortest line connecting two points. It is therefore natural to consider whether precisely this concept of a straight line, with which Kant[11] too was familiar, is the reason why Kant in the one passage says that the concept of a straight biangle is free of contradiction. Indeed, he supports the assertion with the following argument: "The concepts of two straight lines and their meeting contain no denial of a figure."[12] If a straight line is defined as the shortest line connecting two points, then the concept of a "straight biangle" is indeed consistent, since, for example, on the surface of a sphere two lines that are straight in this sense can indeed form a "biangle" as a "figure" that encloses an area.

The problem with this is that in the third passage, which comes just before, Kant himself expressly formulates a "denial" of "figure", which accordingly the concept of a "straight biangle" *does not* contain. He does this by emphasizing: "Two straight lines enclose *no* area"[13], this matching Euclid's Axiom 9. If Kant does appear therefore to have moved away from this position, because the concept of a "straight biangle" is consistent if a straight line is defined as the shortest line connecting two points, then it could well appear that Kant did indeed already consider the possibility of non-Euclidean geometry. For why should not Kant have seen that, for example, on the surface of a sphere a straight biangle is indeed perfectly possible and with it also a non-Euclidean geometry, for which, of course, the geometry on the surface of a sphere gives a model?

The difficulty is that Kant himself could in no way have had this conception, for he did not regard the concept of a "straight biangle" as capable of "construction", whereas it certainly is on the surface of a sphere, being transparently intuitive without any restriction. But it then follows further that Kant in principle could not have conceived the biangle on the surface of a sphere as straight, and therefore could not have allowed a corresponding definition of a straight line as the shortest line connecting two points. Indeed, Kant advanced fundamental criticism of precisely this definition, which, however, has apparently passed unnoticed because Kant, lacking an appropriate definition of his own, was confident in the negative criticism but not able to offer anything positive in its place.

He is critical of the idea that the straight line is allegedly the shortest connection between two points: "My concept of a *straight line* contains nothing of

---

[9] cf. *Elemente*, I, Def. 4

[10] cf., e.g., Archimedes, *Werke*, Darmstadt 1963, p. 78

[11] cf., e.g. B 16

[12] A 220 f. B 268

[13] A 163 B 204, my italics

magnitude but only a quality," whereas in a definition of this kind the expression "shortest" makes "magnitude" or quantity decisive. Kant, however, goes too far when, without having at his disposal a better definition, he believes that he can be certain: "The concept of a shortest line ... cannot be deduced by any analysis from the concept of a straight line. Intuition must therefore be brought to our assistance"[14]. For in saying this he tacitly assumes something that is, not only for his critical philosophy, simply disastrous, namely, that a concept like that of a straight line can be formed *without* intuition and therefore possess definiteness as meaning or sense *without* intuition[15]. It is precisely this assumption that Kant is making when he says in that passage that a straight biangle can, despite its synthetic impossibility, be possible analytically – as though a concept such as that of a straight biangle can possess its meaning even *without* such an intuition as corresponds to the subconcepts "biangle" and "straight".

Therefore, what we must do, standing in for Kant, is to attempt to establish whether concepts like "straight" or "straight line" can after all be satisfactorily defined in precisely the sense that they can originally have only in connection with intuition. For we could then also investigate how they are actually related to other concepts such as "shortest line connecting two points" or "biangle". A definition in Kant's sense could, for example, take the following form: a straight line is a spatially one-dimensional extension that maintains its direction.

Attempts to define a straight line in this way have already often been undertaken, but they probably only failed to establish themselves because they appeared to be circular[16]. Therefore, what we are called upon to do is to demonstrate that here there can be no talk of circularity, and this is something that can be done with the help of philosophy as reflection, which evidently has never been invoked. A definition of this kind suggests itself, since it uses as undefined – given that not everything can be defined – only the concepts that anyone in any way capable of posing the simplest and most rudimentary problem of geometry must always have understood. For the fact that the straight line presents a problem can only be grasped by someone who has always understood that here a straight *line* as *extension* of *one dimension*

---

[14]B 16

[15]Because for Kant to "construct" a concept means not at all to *form* a concept, but to connect a (hence already formed) concept with an intuition which corresponds to it (cf., e.g., A 712 ff. B 740 ff.) Thereby it is the intuition which will be "constructed" as such in the sense of *formed* (cf. Kant himself in A 715 f. B 744 f.), but only namely as the one corresponding *to* a concept. The concept as such would be correspondingly formed thereby without this intuition, although it would remain completely incomprehensible.

[16]cf. H. Schotten, *Inhalt und Methode des planimetrischen Unterrichts*, 2 Vols. Leipzig 1890-93, especially Vol. 2, p. 3 ff., and earlier already H. V. Helmholtz, *Die Tatsachen in der Wahrnehmung*, Darmstadt 1959, p. 58 f. and J. Schultz, *Prüfung der Kantischen "Kritik der reinen Vernunft"*, Königsberg 1789, Part I, p. 58.

in *space* is meant. Further, if it is to be understood as *extension*, it must also be understood that a line *as* extension also has *direction*, because extension as such can only be understood at all as extension in some direction.

However, it is probably precisely because extension is invariably associated with some direction of the extension that the appearance of circularity arises in this kind of definition. For "direction" has no other meaning but "straight direction", so that "curved direction" would from the beginning be only a *contradictio in adjecto*. As a result, the definition of the *straight* line as a spatially one-dimensional extension would, it appears, unavoidably become circular through the use of its "direction" as a "*straight* direction", because "straight" is not thereby defined but much rather presupposed.

However, this appearance arises only when the actual statement of the definition is not properly considered. It in no way states that a straight line is a spatially one-dimensional extension that has a direction but that it is that spatially one- dimensional extension which *maintains* its direction. In this way, direction as *straight* direction is indeed presupposed, but by no means in the sense in which the straight line is to be defined in the first place. For the straight line is not defined through the possession of one-dimensional extension in space and with it direction in general, but rather through the fact that it *maintains* this direction, which it already possesses as a spatially one-dimensional extension in general. It is this *maintaining*, as opposed to the *not* maintaining, which is therefore to be conceived at the same time, that is the special property of the straight line, which is thereby also defined for the first time, so that, conversely, the curved line is also first defined by its special property of the corresponding *not* maintaining.

It follows from this that, in the sense of the direction which the straight line maintains but the curve does not, a 'straight line' as a common possession of both must be distinguished from both, and, like a genus of two species, can neither be in contradiction to the curved line nor identical in meaning to the straight line – and therefore the definition of this straight line through that 'straight line' is also not circular. Only in this way can one explain what geometry takes as directly obvious, namely, that also every (continuous and smooth)[17] curve possesses a direction, which is to be specified geometrically by taking a straight line as tangent to such a curve point for point. For "direction" can here also mean only "straight direction", since it is in principle impossible for a "curved direction" to exist, and therefore a curve, by pos-

---

[17]But also such curves which are continuous but not smooth – in one point or even in all points –, do not at all disprove this *ansatz* of Kant, although this is frequently stated (cf., e.g. M. Friedman, *Kant and the Exact Sciences*, London 1992, p. 78 ff.). Namely, the only positive mark for such a point is that here a curve has infinitely many directions or tangents. Hence it is to be marked positively only by the fact that here a curve changes its direction discontinuously – therefore it possesses basically a direction. Because clearly a curve being continuous must be extension and as such also basically possess a direction.

sessing a *direction*, possesses a *straight* direction and therefore in this sense as curve itself has the straight line within it.

This cannot, however, constitute the contradiction that "locally" such a curve is – even if only "in the infinitely small" – a straight line, as has often been said in the course of "infinitesimal calculus" or "analysis". For also the fact that a straight line is to be associated as tangent with such a curve point for point and the direction of the curve is to be specified through the direction of the tangent does not in any way mean that the curve shares with the tangent a little piece of a straight line but only that it has one point in common with the tangent as straight line. As curve, namely, it has a straight line as tangent, which only *beyond* this point, that is, only as a one-dimensional *extension* in space, is a straight line, as such not somehow within the curve but outside it.

That as a consequence such a curve always contains within it not only its curvature but also with the curvature direction as straight direction can in fact only mean without contradiction that it possesses the straight direction, not as an actual straight line, but as a potential 'straight line', namely, not as species but as genus. This cannot be otherwise if the following is true: like an actual straight line, an actual curve is none other than a quite definite kind of spatial entity, namely, a shape in the form of a space that in all its three dimensions is from the beginning Euclidean, namely, 'straight', as a synthetic a priori fact. But this must be so if it is true that space, of however many dimensions, is of subjective origin, something primordially subjective and *not* something primordially objective, as empiricism assumes without reflection. Although this standpoint was not implemented by Kant, it is still topical, at least in view of the admissions of empiricism: we do not know what space and time are; in particular, we do not know why time has one dimension and space three.

For suppose that space is indeed to be generated synthetically a priori through subjects, that is, according to Kant, is to be extended as extension. Then this extension, as primordial extension of space, can only be 'straight', or uncurved or Euclidean. For considered synthetically a priori, there would be only one reason for extension of space in general and therefore also only one for direction in general and thus for 'straightness', or lack of curvature or Euclidicity. In contrast, any kind of curvature as the deviation from straightness could only appear as such a posteriori giving us empiricism as the set of curved objects of all kinds, through which primordially subjective space becomes for the first time objective space, namely, the space of such objects, in other words, it becomes the form of what is taken into cognizance as the foundation for all empiricism. It would then be necessary for the following to be true too: the attempt to define the straight line leads to a 'straight line' as the genus of the straight line and the curve. Here again no circularity is established; all that is established is a larger context of derivation, which includes this space. For it would also have been shown why any line – whether straight or curved – can be something primordially definite only in the form of 'straight' space.

The fact is that no geometer has yet explained the assumption – nevertheless regularly made – that in terms of quality it is always only a curve that is to be determined as the deviation from a straight line as tangent but never the opposite, a straight line as the deviation from a curve. It is only in terms of quantity that one can, conversely, determine the straight line on the basis of the curve, namely, as the curve with curvature zero. However, the quality of the curve as the deviation from the straight line is here always presupposed. This can only be explained by reflecting that every line is primordially definite only as extension in the form of this Euclidean 'straight' space. It is only in the form of such space that not merely actual curves but also at least geometrically also actual straight lines can be formed: that which always determines actual curves is, after the manner of actual curves, to be actualized, and with its help geometry is to be erected *ab initio* as the general science of spatial definiteness. In accordance with this, actual straight lines are only actualizations of *that* one-dimensional space, which is potentially contained or "embedded" (as geometers say) *in* the three-dimensional space: they are actualizations of the 'straight' one-dimensional space.

For if we disregard two of the dimensions of the 'straight' three-dimensional space by imagining in this sense two sections in order to arrive at a representation of one-dimensional space, this cannot lead to anything but actualization *from* the 'straight' *to* what is straight and one -dimensional, because the three-dimensional space can only be 'straight' already from its first dimension. For this reason, concepts like "straight line" are ultimately only concepts of that which as spatially one-dimensional extension occurs just as actually as curves, by itself actualizing their direction and thereby, in contrast to the curves, maintaining that direction. As something that has definiteness, i.e., sense and meaning, only in so far as it possesses an object, a nonempirical concept like "straight line" can be formed and used only in an indissoluble association with extension in the form of such space as nonempirical pure intuition. Without this, such concepts must therefore lose not only their object but, with it, all definiteness as meaning or sense. Even Hilbert was of this opinion, repeatedly emphasizing in his *Grundlagen der Geometrie* that the formulation of geometrical axioms "amounts to the logical analysis of our spatial intuition" or "expresses certain interconnected basic facts of our intuition"[18].

However, no less a person than Kant always speaks as if a concept can have sense or meaning as definiteness even without intuition. For he is never really clear that the *formation* of concepts, and not just their *application*, is impossible without the corresponding intuition, and this has unfortunate consequences. For Kant could have asked much more decisively whether the geometers were indulging in illusions when they considered themselves free to proceed from the notion of the "straight line" simply as the "shortest line connecting two points"; this, in particular, hinders the determination of the relationship between Euclidean and non-Euclidean geometry. Namely, it is only

---

[18] 13th ed. Stuttgart 1987, p. 1 f.

in the sense of the "shortest line connecting two points' that a "straight line" can exist in non-Euclidean geometry exactly as it can in Euclidean geometry. For the aforementioned reason, Kant did not see, but in the framework of his new conception could have seen, that this notion of the "shortest line connecting two points" by no means exhausts the conceptual content of a "straight line" that is developed in our definition. Kant merely emphasizes that from the concept of "straight line" alone the concept of "shortest connecting line" does not follow[19]; that can come only with the corresponding intuition. However, to emphasize this is both confusing and superfluous, since without such intuition this concept cannot be formed at all. Precisely for this reason he then fails to see that not even with such intuition does the concept "straight line" yield the concept of the "shortest connecting line" in the sense that the latter fully reproduces the former.

In the sense of Kant, a much more important thing would be to recognize that in accordance with our definition every straight line is indeed the shortest connecting line of two points but that the converse – every shortest connecting line of two points is a straight line – is by no means true. Thus, "shortest line connecting two points" is a necessary but in no way sufficient condition for a straight line. The sufficient condition is only provided by our definition, in accordance with which there can be no doubt how we must ultimately resolve Kant's indecision: the concept of a "straight biangle" is without doubt contradictory, because the concept "straight" can possess only the sense that it has together with that intuition.

If one were to fall back to a "straight line" as the "shortest line connecting two points", in order to avoid in this way the contradiction, one would, following Kantian logic, overlook something decisive, namely, one would be presenting this merely necessary condition as sufficient, as is done up to this day for the purpose of developing non-Euclidean geometry. But suppose the non-Euclidean geometries were denied this withdrawal to "straight lines" as the "shortest lines connecting two points"? Would not then the alleged independence and equality of status of the non-Euclidean geometries *vis á vis* Euclidean geometry become at least questionable? Indeed, would we not have stronger ground to ask whether the implementation of the Kantian approach might not prepare a foundation on which the postulate of parallels could still be proved?

It is true that this would not mean, as Frege thought[20], that the non-Euclidean geometries are self-contradictory in the way that the concept of a straight biangle is – that in accordance with our definition the straight biangle could not even exist on the surface of a sphere. But would not this mean that the non-Euclidean geometries – precisely because they are free of contradiction – can no longer, unlike Euclidean geometry, be synthetic a priori geometries

---

[19]cf. again B 16

[20]cf. G. Frege, *Nachgelassene Schriften und Wissenschaftlicher Briefwechsel*, Vol. 1, Hamburg 1969, p. 182 ff.

of synthetic a priori space but only analytical geometries as logically possible formal systems, in the framework of which "space" is then merely metaphor? And would not this also be fully in accord with Einstein's general theory of relativity? For then they could have no bearing on "space" as something nonempirical and a priori, but as empirical non -Euclidean geometries they would merely be concerned with what is "spatial" as something empirical or a posteriori. The empiricist does not distinguish between the two possibilities, because as such he cannot, so that for him space exists only as something emprical and physical, and this space means none other than what is spatial.

Be that as it may, the fact that Kant hesitates on the question of the straight biangle is not at all to be attributed to his having contemplated non -Euclidean geometry for "space" regarded as nonempirical and a priori. Kant's special hesitation whether the concept "straight biangle" is or is not contradictory derives rather from Kant's general hesitation concerning the relationship between concept and intuition. Though Kant is quite certain that only with intuition does a concept possess definiteness as meaning and sense, it is not at all clear to him that then one cannot arise before the other: neither concept before intuition nor intuition before concept. Rather, both must then be equally primary and therefore arise from the unity of the subject as the unity of understanding *and* intuition.

Therefore, it cannot be either that concepts are first formed and then merely "constructed" by intuition and thus referred to an object. For this they would already have to possess, by virtue of their formation and in themselves, definiteness as meaning or sense, and the ultimate consequence of this could only be Platonism. Nor can concepts be formed "analytically"[21] from intuition, still less "abstracted"[22] from an object. The ultimate consequence of that would have to be Aristotelianism, which could not explain how something as specific as a concept could arise through mere "analysis" or "abstraction" from something which in no sense is a concept.

Following Kant, this could only be made comprehensible through primary synthesis, namely, through synthesis of concept and intuition correlated together. The reason why nevertheless Kant continually hesitates between these, for him, untenable possibilities of Platonism or Aristotelianism is simply because he did not gain insight into this primary synthesis of the unity of understanding and intuition as unity of the subject. Only in this way could he have shown that in this respect Aristotle and Plato represented false alternatives to each other and that the correct alternative could only be an alternative to both.

---

[21]cf., e.g., A 76 f. B 102 f., A 78 f. B 104 f., B 133 f. with commentary
[22]so especially in the *Logik*, cf. Akademieausgabe Vol. 9, p. 91 ff.

Substance as Function: Ernst Cassirer's Interpretation of Leibniz as Criticism of Kant[1]

Enno Rudolph

FESt, Schmeilweg 5, D-69118 Heidelberg, Germany

It cannot now be denied that Cassirer distanced himself significantly – not only in the philosophy of science – from the eponym of his school: Neo-Kantianism. The question therefore is only how far he distanced himself – did he become so original that one can no longer call him a Kantian, or did he merely skilfully exploit the hermeneutic latitude that Kant himself allowed without crossing the bounds of transcendental criticism. This question can be discussed on the basis of numerous individual fields of Cassirerian philosophy – philosophy of science is just one of them. It seems to me that from the confrontation between the epistemological requirements that Cassirer derives in his book "Determinismus und Indeterminismus in der modernen Physik" and epistemology in Kant's sense there follows a difference from Kant's position that can be explained by the choice of a theory of concept formation that Cassirer already made and elucidated in 1902 in connection with his interpretation of Leibniz's philosophy, and to which he remained true. Confirmation of this observation would not only be of interest biographically for the development of Cassirer's thought; more, it could help us to establish which standpoint in the philosophy of science in the post-Renaissance period can be appropriately compared with the conceptual reflections of modern physics: Kant's apriorism of the pure concepts as the golden mean between the sceptical extreme of David Hume, on the one hand, and the substantialistic extreme of Leibniz, on the other, or the theory of order-creating individual concepts (*notions individuelles*), which was developed by the same Leibniz and, if one follows Cassirer, is not necessarily to be interpreted in the sense of an absolute substantialism. It appears to me that examination of the manner of Cassirer's representation of the topicality of Leibniz's philosophy is a suitable means to trace the route that he, employing Kantian notions, embarked on, leaving behind not only the epistemological orthodoxy of pure apriorism but also every form of neo-Cartesian ontological duality between subject and object.

---

[1] translated from German by Julian Barbour, Oxon, UK.

I

One way of reading Cassirer's book on Leibniz is as a study of the modification that the concept of substance underwent at the hands of Leibniz in reacting, on the one hand, to his intimate foe Descartes and on the other to his ontological archpriest Aristotle. In this point, Cassirer is strikingly unambiguous and determined. The analogous relation between substance (as carrier of attributes) and attributes in the sense of the Leibnizian logic of *predicatum inest subjecto* and the metaphysics of monads as essential inclusion of all their virtual states occurs in the framework of the natural philosophical writings of Leibniz in the context of his doctrine of force. It is force as quality that creates and, in its different quantifiable (measurable) states, sustains itself – the *vis primitiva activa* [2]– that here represents substance, and which Cassirer attempts to keep apart from every reifying misunderstanding. For him, the continuous "persistence" – the decisive characteristic of "substance" – of force is not to be understood as "material constancy of a thing" but as "the persisting identity of a superordinate law" [3]. Cassirer elucidates in another place the concept of substance in the same way.

In order to meet the Leibnizian requirement of a well-ordered relationality of the individual units of a complete process – a requirement that holds just as well for the biography of a cognizant monad, i.e., a human being, as for free fall or the rolling of the famous ball on an inclined plane – one must assume that each individual member of a series, i.e., every individual well-defined state of a natural phenomenon, each instant of its development in time, is a particular case of the ordering principle of the complete process, and is therefore to be understood as a particularization, or instantiation, of a "system". "The primitive force," concludes Cassirer, "is nothing else but this system, this unitary quintessence [einheitliche Inbegriff] of the different force states for the complete temporal series" [4]. In what follows, I wish to argue that Cassirer's tendency throughout is to show it is this sense of essence that for him is physically more relevant than a priori, and thereby quasisubstantial, concepts, and that this type of essence avoids the suspicion of being merely equivalent to substance because Cassirer, who here indeed is literally Kantian, understands this quintessence, not as preexistent, but as grasped hermeneutically.

In the Leibnizian scheme, the forces that we measure, the "derivative forces", i.e., the "individual members" in the above sense, are, in Cassirer's interpretation, to be seen as the results of antecedent changes if we wish to be able to measure them at all – measuring is also a manner of understanding – and it is the "system" that determines these results and therefore enables them to 'represent' the complete system, just as the monads represent the world. "The determination of the state is therefore to be conceived as analogous to

---

[2] Leibniz (1982), p. 7
[3] Cassirer (1980), p. 300f.
[4] Cassirer (1980), p. 301

the ideal characterization of the point of a curve by the direction of the tangent, whereas the primitive force corresponds to the law of continuous change of direction, through which the actual orbit of the point in fact arises" (loc. cit.). Cassirer would later have replaced the vague talk of "correspondence" between the *vis primitiva activa* and the "law" by the word symbolization. However, more important in our connection than this readily verified observation is the way in which Cassirer already sees, in this correlation established by Leibniz between substantial force, on the one hand, and law, on the other, the preparation – in part indeed the implementation – of Leibniz's passage from the substance concept (in the sense of the "constancy of a thing") to the function concept. With regard to the force, he says that "mathematically speaking," it contains "the lawfulness of the function, to the extent that this is regarded as being determined by the totality of the higher derivatives at a point, whereas the logical sense of the derivative force is already exhausted by the first derivative" (loc. cit.). Cassirer speaks here of "function" in a sense that has already been established. He explicitly credits Leibniz with a "logical generalization" of the concept of function, in accordance with which the function concept will have the general meaning of "mutual logical dependence of thought contents" [5]. Making this more precise, Cassirer adds that by "dependence" one is here to understand that "one content of thought arises from another lawfully and with unique definiteness" (loc. cit.).

What is said here, formally, about the logical significance of the function concept evidently applies correspondingly to the physical significance in the above sense of the "lawfulness of a function" and also the mathematical significance, of which it is said explicitly that the aim of this establishment of the function concept is that "...the entire manifold of spatial shapes is to be built up out of the point" (loc. cit.).

Summarizing, we can say that in Cassirer's interpretation of Leibniz a law, understood as the determining essence of all possible attributes or all possible states for a succession of phenomena in time (or for the relationships between phenomena in space), takes over the role of substance, in the same way that a mathematical function is to be regarded accordingly as the essence of all possible points in their relationship to other points on, for example, a curve. Thus, substance is to be understood as the "quintessence" [Inbegriff] of relationality.

---

[5] Cassirer (1980), p. 148

II

A concept that fulfils this quintessential [inbegriffliche] performance is, through its richness of characteristics, superior in accuracy to the type of concept that is won through a 'normal' abstraction process, i.e., is of the traditional kind, "that something general, by virtue of the uniformity with which its content recurs in the changing particular, stands out on its own account ..." [6]. In fact, Cassirer has no intention of renouncing this abstracting capacity, but he regards it at best as the *conditio sine qua non* for the fulfilment of the epistemologically relevant requirement on concepts that they be appropriately symbolized by individual perceptions. The same is true of the logical, respectively mathematical, requirement on concepts that they have an appropriate symbolic representation through particular properties, respectively through individual points of a line (say).

The requirement already met by Leibniz, namely, the replacement of substantial constancy of some uniform general thing by the *notion individuelle*, understood as "quintessence" [Inbegriff], provides Cassirer with the material for his criticism of the inadequacy of the traditional abstraction procedure. He also calls it "negative abstraction", which is said to be in constant danger of "destroying everything definite" [7] because the degree of abstraction – precisely for the sake of the general validity of the concept (Cassirer illustrates this for the example of the concept "metal" [8]) – makes it impossible to find any "return to the concrete special cases." Cassirer, who appeals to Lotze's logic in his criticism of the inadequacy of "negative abstraction", contrasts this method with a model of "positive abstraction", which should counter the "stunting" associated with every conceptual generalization in that it helps to purchase the ticket for the desired return journey. De facto, Leibniz's "essence" [Inbegriff] thereby achieves new honours: "The fixed properties [i.e., of the concepts won in the negative abstraction process] are replaced by general rules that enable us to survey in a glance a complete series of possible determinations" [9].

Cassirer concentrates on two aspects of the essence won by "positive abstraction" that demonstrate its power, without exhausting himself with arguments from thought-psychology, which he certainly values [10]:

1. The knowledge of uniformity – the characteristic of a concept won by "negative abstraction" – is augmented and in certain cases corrected by a *knowledge that certain individual elements belong together*. Agreeing with Benno Erdmann, Cassirer argues here that a new phenomenon, a new nuance in a known series of diverse colours or a new chemical compound, does not at all need to be repeated in order to adjoin it, the name or the compound.

---

[6] Cassirer (1990), p. 30

[7] Cassirer (1990), p. 28

[8] cf. Cassirer (1990), p. 27f.

[9] Cassirer (1990), p. 29

[10] cf. Cassirer (1990), p. 23

The element is fixed ("recognized") as a "member of the series"; at the same time, it is taken into the order of the relationality as belonging to the series. In other words: the elements from which a concept is distilled by negative abstraction are only "legible" as elements capable of conceptualization – who is here not reminded of Kant's "spelling out the phenomena" – because these elements are from the beginning conceived as already connected by a definite relation. Cassirer is here accordingly adopting the thesis that thought never encounters isolated elements but always meets them in the context of an order that we continually renew or augment. This is an order of relationship between the elements that is always greater than the individual elements that are recognized and always smaller than a presupposed absolute horizon of the world of experience. At the first encounter, this appears as a pleading for the apriorism of the pure concept (as condition of the possibility of experience in the sense of Kant) if not for a modern version of the doctrine of anamnesis. Looked at carefully, above all when Cassirer is read backwards, it turns out that Cassirer extends this requirement in the epistemologically relevant part of his discussions to all insinuation of preconceptual conditions of experience, that is, he does not allow any independent thing, on which the concept must first prove itself, to stand opposite the aprioricity of thought, and vice versa he does not allow any concept that, even empty, can be conceived without intuition. In possible phenomena, however amorphous they may be, the essence always symbolizes itself; otherwise they are "for us nothing," even when we "forget" this. Cassirer knows no thing in itself ("Ding an sich").

2. Cassirer's separation from Kant goes even further when one includes the second characteristic of the essence: *the creative aspect.* "The significance of the *law* which connects the individual elements is not exhausted by the enumeration of however many *cases* of the law; for this enumeration would lack precisely the creating *principle* that makes the individual elements capable of connection to a functional essence. If I know the relation through which $abc...$ are ordered, I can, by reflection, consider them individually and make them separate objects of thought; in contrast, it is impossible from the bare presence together of a, b, c to win a notion of the nature of the connecting relation .... This conception cannot run the danger of reifying the pure concept, of according it an independent *reality* alongside the individual things"[11]. Indeed: Cassirer remained true continuously to the belief in relationality as the quintessence of nature. It occurs in the Leibniz study as the thought that the relation is the mirror of the *lex seriei*, in SF as the ordering element of an interconnection nexus that replaces the mere commonality of characteristics as established in "negative abstraction", and finally in D/I, where we have the fundamental statement: "In the strict physical use of language, 'nature' is nothing but a quintessence of relations, of laws – and," adds Cassirer refusing of the type of dualistic relation as characterized by Kant in the form of the law establishing unconditional subject, on the one hand, and "obedient" nature,

---

[11] Cassirer (1990), p. 33f.

on the other, "to such an essence, to pure 'form' of such kind, the category of action or suffering cannot be applied" [12].

Cassirer sees this concept of nature explicitly confirmed by the development of modern physics, and with almost melancholy regret he notes after the event that in SF he could only demonstrate this on the basis of the state of development of classical physics: "I have attempted in another place to show in detail how this 'substantial' conception is gradually transformed, how the substance concept is more and more displaced and replaced by the pure *function concept*. In these considerations, I restricted myself to the development and the problems of classical physics; however, they could have been presented in a much more succinct and pregnant form had the general theory of relativity and modern atomic physics existed at the time I formulated them. Today this development of thought has been brought to a secure and, apparently, definitive conclusion." [13].

Naturally, one question remains open, namely, that of the manner in which, according to Cassirer, the "creation" (of the whole out of the part), which is an integral aspect of the Leibnizian model, is to be conceived in a "desubstantialized" nature. It appears not only that Cassirer thought the idea of creation of orders on the basis of a "quintessence" – orders whose meaning finds its representation in the formulation of the law – was analogous to the manner in which the Leibnizian *substance individuelle* brings forth reality, but even that he believed the creative aspect to be already exhausted in such an establishment of order recognition. Several things support this thesis of Cassirer, for example, that one could win in this manner already in Leibniz's writings a plausible equivalence, without metaphysical surplus, between *notion individuelle* and *substance individuelle*. The persistence of the creating principle is the valid law, and to seek a reality "beyond" the law that comes before the law as "thing in itself" is senseless and bad metaphysics, because "we cannot speak of physical 'existence' otherwise than under the conditions of physical cognition, both its general conditions and the particular conditions that hold for its observations and measurements" [14].

The independent existence that appears to stand opposite the cognition process in an unfortunate relation of object and subject can at best, in Cassirer's view, be conceived as a limit to which the cognition process "tends" (loc. cit.). "We do not simply read off the laws from the objects but condense the empirical data that are accessible to us through observation and measurement to laws and thus to objective statements, and outside this there does not exist for us any objective reality that we should study and seek" (loc. cit.).

If we take an epistemological concept restricted in this manner as not only physical but also philosophical, then the conclusion that it is meaningless to seek an existence outside the cognition process should also be applied to the

---

[12]Cassirer (1957), p. 263
[13]Cassirer (1957), p. 277
[14]Cassirer (1957), p. 279

object of cognition. It can no longer be taken as the unconditional prerequisite for the setting up of laws as conceived in the model of the "reading off of laws" or "spelling out the phenomena" in the Kantian sense. Instead, the "process of objectivization" is to be conceived without alpha, without omega, without substance, and – *horrible dictu* – without subject. To the extent that one can speak of laws as "conditions for the possibility of experential knowledge" [Erfahrungserkenntnis], one may still call this method "transcendental", but nothing decisive depends on this. However, the proud claim to the unconditional aprioricity of a "transcendental subject" must give way to the modest requirement of a hypothetical supposition that employs continuously verified laws as proved means, which, however, must be corrected when they are no longer confirmed.

The process of objectivization itself is, in addition, to be reconstructed as a constituent process of such a story, which can be subdivided into a series of other constituent processes – processes like mythical, artistic or linguistic symbolizations, etc., that may unfold in succession or with their phases intertwined and make up a nexus of phenomena that Cassirer calls culture. If there is no existence beyond the cognition process, if rather nature exists only as an "essence of relations, of laws" – and not of objects to the extent that they are subject to laws (Kant) – then there can be no ontological distinction between nature and culture. In so far as Cassirer refuses both culture-independent and culture-conditioned meaning just as he abjures the idea of a substance that carries the laws of nature, he is neither a metaphysician nor even a "metaphysician of metaphysics".

The distancing from Kantianism undertaken with Kantian means culminates in fact in a remarkably heretical reinterpretation of the cognition-guaranteeing rank of the Kantian categories, exquisitely demonstrated for the model physical example: causality. "If one wishes to express it in the language of Kant," the law of causality belongs, according to Cassirer, "to the modal principles: It is a 'postulate of empirical thought'" [15]. That is precisely what it is not for Kant. For Kant it belongs to the "analogies of experience" – and is thus to a high degree prescriptive, determining, and "anticipatory" as the modal modesty of the postulate of empirical thought. The elucidation by Cassirer of a mere "postulate" nature of the law of causality fully confirms his departure from the apriorism of pure understanding in its pure anticipatory function. According to Cassirer, postulate means essentially nothing else than that "the process of conversion of the observational data into exact statements of measure, the gathering together of the measurement results in equations between functions, and the systematic unification of these equations through general principles" can never be completed (loc. cit.).

The "danger" that arises for the physical concept of object from this view of things is named explicitly by Cassirer [16]. In SF, he had already indicated

---

[15] Cassirer (1957), p. 197
[16] cf. Cassirer (1957), p. 285

how one was to handle it: The ordering concept (causality), or the law that represents it, serves as a sign that points to a "quintessence" [Inbegriff], and in this sense "denotes" it. Perceptions can only be "understood" at all – and all cognition is also understanding – if we interpret them as signs. However, we can only interpret signs under the assumption of an order of relation between the signs. Signs do not represent the relation between our knowledge but the quintessence of the perceptions set as a task. When in another place it is said that we do not "think objects" [Gegenstände denken] but "think objectively" [gegenständlich denken], here one could also say that we do not think "conceptually" [begrifflich] but "quintessentially" [inbegrifflich]. In this modest arena, and only thus, do we come to something like cognition, and it must be continuously reformulated when the current ordering models, the laws, no longer prove viable.

It is at the most the farewell to divine knowledge as the instance of sufficient knowledge of the "quintessence" of all relations that divides Cassirer from Leibniz and that also only if Leibniz meant by the urmonad (monas monadum) more than the fictitious paradigm of monad in general as an absolute substance whose ontological rank consists in guaranteeing real existence to every natural object: "If, thus, one defines the object, not as an absolute substance beyond all cognition, but as an object that shapes itself in developing experience, then there is here no 'epistemological chasm' that has to be overcome laboriously by some command of thought, by a 'transsubjective injunction'" [17].

References

Cassirer, Ernst (³1980): Leibniz's System in seinen wissenschaftlichen Grundlagen, Hildesheim, New York.

Cassirer, Ernst (⁶1990): Substanzbegriff und Funktionsbegriff. Untersuchungen über die Grundfragen der Erkenntniskritik, (abbreviated: SF).

Cassirer, Ernst (⁵1957): Determinismus und Indeterminismus in der modernen Physik. In: Cassirer, Ernst, Zur modernen Physik, Darmstadt (abbreviated: D/I).

Leibnitz, Gottfried Wilhelm: Specimen Dynamicum, hg., bers. u. kommentiert v. H.G. Dosch, G.W. Most u. E. Rudolph, Hamburg 1982 (abbreviated: Spec. Dyn.)

---

[17]Cassirer (1990), p. 395

Forthcoming

U. Ratsch, E. Rudolph and I. O. Stamatescu (Eds.)

Philosophy, Theology and Modern Physics

Background Questions of a Discourse

Contents:

J. Hübner: *Kosmologie zwischen Mythos und Logos.*

G. Börner: *Das Standardmodell der Kosmologie.*

G. Raffelt: *Teilchenphysik und Kosmologie.*

M.G. Schmidt: *Superstringtheorie.*

H.D. Zeh: *Warum Quantenkosmologie?*

T. Görnitz: *Relations between Universal Quantum Theory and Space-Time.*

E. Rudolph: *Weltentstehungsmythos bei Platon; Anmerkungen zur platonischen Kosmologie.*

J. Seidengart: *Infinitistische Kosmologie und Kopernikanismus bei Giordano Bruno.*

H. Wismann: *Mathematics and Physical Reality in the Philosophy of Plato.*

The authors of this issue of "Texte und Materialien" participated in the conferences on Physics and Philosophy where the papers included in the book "Philosophy, Mathematics and Modern Physics, A Dialogue" were originally presented. Their contributions supplement the discussion published in the volume at hand. They add in particular to the argumentation concerning space-time and cosmology.

Texte und Materialien, Reihe A, Spring 1994
Available from: Forschungsstätte der Evangelischen Studiengemeinschaft, Schmeilweg 5
D-69118 Heidelberg, Tel: 06221/9122-0, Fax: 06221/167257

Springer-Verlag
and the Environment

We at Springer-Verlag firmly believe that an international science publisher has a special obligation to the environment, and our corporate policies consistently reflect this conviction.

We also expect our business partners – paper mills, printers, packaging manufacturers, etc. – to commit themselves to using environmentally friendly materials and production processes.

The paper in this book is made from low- or no-chlorine pulp and is acid free, in conformance with international standards for paper permanency.